Engineering Analysis COSMOSWorks Professional 2007

Paul M. Kurowski

Mr Ravi Bhutoria
Apt NO. N M108,
1919 University Dr NW
Calgary, AB T2N 4L5
creating a future without breast cancer

ISBN: 978-1-58503-353-9

PUBLICATIONS

Schroff Development Corporation

www.schroff.com
www.schroff-europe.com

Trademarks and Disclaimer

SolidWorks and its family of products are registered trademarks of Dassault Systèmes. COSMOSWorks is registered trademarks of Structural Research & Analysis Corporation. Microsoft Windows and its family products are registered trademarks of the Microsoft Corporation.

Every effort has been made to provide an accurate text. The author and the manufacturers shall not be held liable for any parts developed with this book or held responsible for any inaccuracies or errors that appear in the book.

Copyright © 2007 by Paul M. Kurowski

All rights reserved. This document may not be copied, photocopied, reproduced, transmitted, or translated in any form or for any purpose without the express written consent of the publisher, Schroff Development Corporation.

Examination Copies:

Books received as examination copies are for review purposes only and may not be made available for student use. Resale of examination copies is prohibited.

Electronic Files:

Any electronic files associated with this book are licensed to the original user only. These files may not be transferred to any other party.

Acknowledgements

Writing this book was a substantial effort that would not have been possible without the help and support of my professional colleagues. I would like to thank:

Maciej J. Kurowski — Mathseed Expeditions Tutoring

Suchit Jain — SolidWorks Corporation

I would like to thank the students attending my various training courses in Finite Element Analysis, for their questions and comments that helped to shape the unique approach this book takes. I thank my wife Elzbieta for her support and encouragement that made it possible to write this book.

About the Author

Dr. Paul Kurowski obtained his M.Sc. and Ph.D. in Applied Mechanics from Warsaw Technical University. He completed postdoctoral work at Kyoto University. Dr. Kurowski is a professor in the Department of Mechanical and Materials Engineering, at the University of Western Ontario. His teaching experience includes Finite Element Analysis, Machine Design, Mechanics of Materials, Kinematics and Dynamics of Machines, Product Development, and Solid Modeling. Dr. Kurowski is also the President of Design Generator Inc., a consulting firm with expertise in Product Development, Design Analysis, and training in Computer Aided Engineering methods. His interests focus on Computer Aided Engineering methods used as tools of product design.

Dr. Kurowski has published many technical papers and taught professional development seminars for the Society of Automotive Engineers (SAE), the American Society of Mechanical Engineers (ASME), the Association of Professional Engineers of Ontario (PEO), the Parametric Technology Corporation (PTC), Rand Worldwide, SolidWorks Corporation and others.

Dr. Kurowski is a member of the Association of Professional Engineers of Ontario and the Society of Automotive Engineers. He can be contacted at www.designgenerator.com

About the cover

The image on the cover presents the deformation plot of the third mode of vibration of a tank. See Chapter 20 for details.

Table of Contents

Before You Start   1

    Notes on hands-on exercises

    Prerequisites

    Selected terminology

1: Introduction   3

    What is Finite Element Analysis?

    Who should use Finite Element Analysis?

    Objectives of FEA for Design Engineers

    What is COSMOSWorks?

    Fundamental steps in an FEA project

    Errors in FEA

    A closer look at finite elements

    What is calculated in FEA?

    How to interpret FEA results

    Units of measure

    Using on-line help

    Limitations of COSMOSWorks Professional

2: Static analysis of a plate   25

    Using COSMOSWorks interface

    Linear static analysis with solid elements

    The influence of mesh density on results

    Finding reaction forces

    Controlling discretization errors by the convergence process

    Presenting FEA results in desired format

3: Static analysis of an L-bracket — 63
Stress singularities
Differences between modeling errors and discretization errors
Using mesh controls
Analysis in different SolidWorks configurations

4: Stress and frequency analysis of a thin plate — 77
Use of shell elements for analysis of thin walled structures
Frequency analysis

5: Static analysis of a link — 91
Symmetry boundary conditions
Defining restraints in a local coordinate system
Preventing rigid body motions
Limitations of small displacements theory

6: Frequency analysis of a tuning fork — 99
Frequency analysis with and without supports
Rigid body modes
The role of supports in frequency analysis

7: Thermal analysis of a pipeline component and a heater — 105
Steady state thermal analysis
Analogies between structural and thermal analysis
Analysis of temperature distribution and heat flux

8: Thermal analysis of a heat sink — 117
Analysis of an assembly
Global and local Contact/Gaps conditions
Steady state thermal analysis
Transient thermal analysis
Thermal resistance layer
Use of section views in results plots

9: Static analysis of a hanger — 133
Analysis of assembly
Global and local Contact/Gaps conditions
Hierarchy of Contact/Gaps conditions

10: Analysis of contact stress between two plates — 145
Assembly analysis with surface contact conditions
Contact stress analysis
Avoiding rigid body modes

11: Thermal stress analysis of a bi-metal beam — 151
Thermal stress analysis of an assembly
Use of various techniques in defining restraints
Shear stress analysis

12: Buckling analysis of an L-beam — 159
Buckling analysis
Buckling load safety factor
Stress safety factor

13: Design optimization of a plate in tension — 163
Structural optimization analysis
Optimization goal
Optimization constraints
Design variables

14: Static analysis of a bracket using adaptive solution methods — 173

P-elements

P-adaptive solution method

Comparison of h-elements and p-elements

15: Design sensitivity analysis of hinge supported beam — 183

Design sensitivity analysis using Design Scenario

16: Drop test of a coffee mug — 191

Drop test analysis

Stress wave propagation

Direct time integration solution

17: Selected large displacements problems — 199

Large displacements analysis

Creating a shell element mesh on the face of a solid

18: Mixed meshing problem — 213

Using solid and shell elements in the same mesh

19: Static analysis of a weldment using beam elements — 217

Comparison between solid, shell and beam elements

Using beam elements for analysis of a weldment

20: Miscellaneous topics — 233

Selecting the automesher

Mesh quality

Solvers and solvers options

Displaying mesh in result plots

Automatic reports

E drawings

Non-uniform loads

Bearing load

Frequency analysis with pre-stress
Shrink fit analysis
Rigid connector
Pin connector
Bolt connector

21: Implementation of FEA into the design process   253
FEA driven design process
FEA project management
FEA project checkpoints
FEA report

22: Glossary of terms   261

23: Resources available to FEA Users   269

Before You Start

Notes on hands-on exercises

This book goes beyond a standard software manual because its unique approach concurrently introduces you to COSMOSWorks software and the fundamentals of Finite Element Analysis (FEA) through hands-on exercises. We recommend that you study the exercises in the order presented in the book. As you go through the exercises, you will notice that explanations and steps described in detail in earlier exercises are not repeated later. Each subsequent exercise assumes familiarity with software functions discussed in previous exercises and builds on the skills, experience, and understanding gained from previously presented problems. Exceptions to the above are Chapters 20-23 which do not include hands-on exercises.

The functionality of COSMOSWorks 2007 depends on which software bundle is used. In this book we cover the functionality of COSMOSWorks Professional 2007. Functionality of different bundles is explained in the following table:

COSMOSWorks Designer	COSMOSWorks Professional	COSMOSWorks Advanced Professional
Linear static analysis of parts and assemblies	The features of COSMOSWorks Designer plus: Frequency (modal) analysis Buckling analysis Drop test analysis Thermal analysis Fatigue analysis * Optimization analysis *	The features of COSMOSWorks Professional plus: Nonlinear analysis** Dynamic analysis Composite analysis

* Fatigue and optimization analyses are not covered in this book.

** COSMOSWorks Professional has some Nonlinear analysis capabilities; they will be illustrated in this book

All exercises use SolidWorks models, which can be downloaded from http://www.schroff.com/resources.

This book is not intended to replace regular software manuals. While you are guided through the specific exercises, not all of the software functions are explained. We encourage you to explore each exercise beyond its description by investigating other options, other menu choices, and other ways to present results. You will soon discover that the same simple logic applies to all functions in COSMOSWorks software.

Prerequisites

We assume that you have the following prerequisites:

- An understanding of Mechanics of Materials
- Experience with parametric, solid modeling using SolidWorks software
- Familiarity with the Windows Operating System

Selected terminology

The mouse pointer plays a very important role in executing various commands and providing user feedback. The mouse pointer is used to execute commands, select geometry, and invoke pop-up menus. We use Windows terminology when referring to mouse-pointer actions.

Item	Description
Click	Self explanatory
Double-click	Self explanatory
Click-inside	Click the left mouse button. Wait a second, and then click the left mouse button inside the pop-up menu or text box. Use this technique to modify the names of folders and icons in COSMOSWorks Manager.
Drag	Use the mouse to point to an object. Press and hold the left mouse button down. Move the mouse pointer to a new location. Release the left mouse button.
Right-click	Click the right mouse button. A pop-up menu is displayed. Use the left mouse button to select desired menu command.

All SolidWorks files names appear in CAPITAL letters, even though the actual file name may use a combination of small and capital letters. Selected menu items and COSMOSWorks commands appear in **bold**, SolidWorks configurations, folder, and icon names appear in *italics* except in captions and comments to illustrations.

1: Introduction

What is Finite Element Analysis?

Finite Element Analysis, commonly called FEA, is a method of numerical analysis. FEA is used for solving problems in many engineering disciplines such as machine design, acoustics, electromagnetism, soil mechanics, fluid dynamics, and many others. In mathematical terms, FEA is a numerical technique used for solving field problems described by a set of partial differential equations.

In mechanical engineering, FEA is widely used for solving structural, vibration, and thermal problems. However, FEA is not the only available tool of numerical analysis. Other numerical methods include the Finite Difference Method, the Boundary Element Method, and the Finite Volumes Method to mention just a few. However, due to its versatility and high numerical efficiency, FEA has come to dominate the engineering analysis software market, while other methods have been relegated to niche applications. When implemented into modern commercial software, both FEA theory and numerical problem formulation become completely transparent to users.

Finite Element Analysis used by Design Engineers

As a powerful tool for engineering analysis, FEA is used to solve problems ranging from very simple to very complex. Design engineers use FEA during the product development process to analyze the design-in-progress. Time constraints and limited availability of product data call for many simplifications of their analysis models. At the other end of scale, specialized analysts implement FEA to solve very advanced problems, such as vehicle crash dynamics, hydro forming, and air bag deployment.

This book focuses on how design engineers use FEA implemented in COSMOSWorks as a design tool. Therefore, we highlight the most essential characteristics of FEA as performed by design engineers as opposed to those typical for FEA preformed by analysts.

FEA for Design Engineers: another design tool

For design engineers, FEA is one of many design tools that are used in the design process and include CAD, prototypes, spreadsheets, catalogs, hand calculations, text books, etc.

FEA for Design Engineers: based on CAD models

Modern design is conducted using CAD, so a CAD model is the starting point for analysis. Since CAD models are used for describing geometric information for FEA, it is essential to understand how to prepare CAD geometry in order to produce correct FEA results, and how a CAD model is different from an FEA model. This will be discussed in later chapters.

FEA for Design Engineers: concurrent with the design process

Since FEA is a design tool, it should be used concurrently with the design process. It should drive the design process rather than follow it.

Limitations of FEA for Design Engineers

As you can see, FEA used in the design environment must meet high requirements. An obvious question arises: would it be better to have a dedicated specialist perform FEA and let design engineers do what they do best – design new products? The answer depends on the size of the business, type of products, company organization and culture, and many other tangible and intangible factors. A general consensus is that design engineers should handle relatively simple types of analysis, but do it quickly and of course reliably. Analyses that are very complex and time consuming cannot be executed concurrently with the design process, and are usually better handled either by a dedicated analyst or contracted out to specialized consultants.

Objectives of FEA for Design Engineers

The ultimate objective of using the FEA as a design tool is to change the design process from repetitive cycles of "design, prototype, test" into a streamlined process where prototypes are not used as design tools and are only needed for final design verification. With the use of FEA, design iterations are moved from the physical space of prototyping and testing into the virtual space of computer simulations (Figure 1-1).

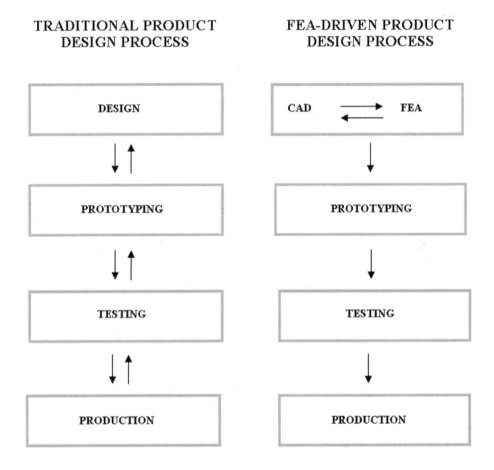

Figure 1-1: Traditional and. FEA-driven product development

Traditional product development needs prototypes to support design in progress. The process in FEA-driven product development uses numerical models, rather than physical prototypes to drive development. In an FEA-driven product, the prototype is no longer a part of the iterative design loop.

What is COSMOSWorks?

COSMOSWorks is a commercial implementation of FEA capable of solving problems commonly found in design engineering, such as the analysis of displacements, stresses, natural frequencies, buckling, heat flow, etc. It belongs to the family of engineering analysis software products developed by the Structural Research & Analysis Corporation (SRAC). SRAC was established in 1982 and since its inception has contributed to innovations that have had a significant impact on the evolution of FEA. In 1995 SRAC partnered with the SolidWorks Corporation and created COSMOSWorks, one of the first SolidWorks Gold Products, which became the top-selling analysis solution for SolidWorks Corporation. The commercial success of COSMOSWorks integrated with SolidWorks CAD software resulted in the acquisition of SRAC in 2001 by Dassault Systèmes, parent of SolidWorks Corporation. In 2003, SRAC operations merged with SolidWorks Corporation.

COSMOSWorks is integrated with SolidWorks CAD software and uses SolidWorks for creating and editing model geometry. SolidWorks is a solid, parametric, feature-driven CAD system developed specifically for the Windows operating system. Many other CAD and FEA programs were originally developed in a UNIX environment and only later ported to Windows, and therefore are less integrated with Windows than SolidWorks and COSMOS.

Fundamental steps in an FEA project

The starting point for any COSMOSWorks project is a SolidWorks model, which can be a part or an assembly. First, material properties, loads, and restraints are defined. Next, as always the case with using any FEA-based analysis tool, the model geometry is split into relatively small and simply shaped entities called finite elements. The elements are called "finite" to emphasize the fact that they are not infinitesimally small, but relatively small in comparison to the overall model size. Creating finite elements is commonly called meshing. When working with finite elements, the COSMOSWorks solver approximates the sought solution (for example stress) by assembling the solutions for individual elements.

From the perspective of FEA software, each application of FEA requires three steps:

- Preprocessing of the FEA model, which involves defining the model and then splitting it into finite elements
- Solving for wanted results
- Post-processing for results analysis

We will follow the above three steps in every exercise.

From the perspective of FEA methodology, we can list the following FEA steps:

- Building the mathematical model
- Building the finite element model by discretizing the mathematical model
- Solving the finite element model
- Analyzing the results

The following subsections discuss these four steps.

Building the mathematical model

The starting point to analysis with COSMOSWorks is a SolidWorks model. Geometry of the model needs to be meshable into a correct finite element mesh. This requirement of meshability has very important implications. We need to ensure that the CAD geometry will indeed mesh and that the produced mesh will provide the data of interest (e.g. stresses or temperature distribution) with acceptable accuracy.

The necessity to mesh often requires modifications to the CAD geometry, which can take the form of defeaturing, idealization, and/or clean-up:

Term	Description
Defeaturing	The process of removing geometry features deemed insignificant for analysis, such as external fillets, chamfers, logos, etc.
Idealization	A more aggressive exercise that may depart from solid CAD geometry by, for example, representing thin walls with surfaces and beams with lines.
Clean-up	Sometimes needed because geometry must satisfy high quality requirements to be meshable. To cleanup, we can use CAD quality-control tools to check for problems like sliver faces, multiple entities, etc. that could be tolerated in the CAD model, but would make subsequent meshing difficult or impossible.

It is important to mention that we do not always simplify the CAD model with the sole objective of making it meshable. Often we must simplify a model even though it would mesh correctly "as is", but the resulting mesh would be too large (in terms of the number of elements) and consequently, the meshing and the analysis would take too long. Geometry modifications allow for a simpler mesh and shorter meshing and computing times.

Sometimes, geometry preparation may not be required at all; successful meshing depends as much on the quality of geometry submitted for meshing as it does on the capabilities of the meshing tools implemented in the FEA software.

Having prepared a meshable, but not yet meshed geometry, we now define material properties (these can also be imported from a CAD model), loads and restraints, and provide information on the type of analysis that we wish to perform. This procedure completes the creation of the mathematical model (Figure 1-2). Notice that the process of creating the mathematical model is not FEA specific. FEA has not yet entered the picture.

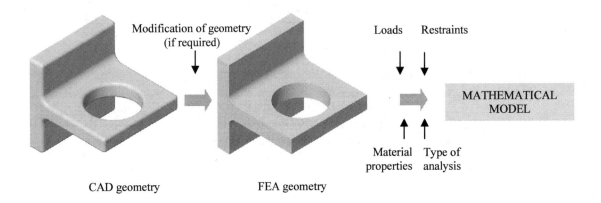

Figure 1-2: Building the mathematical model

The process of creating a mathematical model consists of the modification of CAD geometry (here removing external fillets), definition of loads, restraints, material properties, and definition of the type of analysis (for example, static) that we wish to perform.

Building the finite element model

The mathematical model now needs to be split into finite elements in the process of discretization, more commonly known as meshing (Figure 1-3). Geometry, loads, and restraints are all discretized. The discretized loads and restraints are applied to the nodes of the finite element mesh.

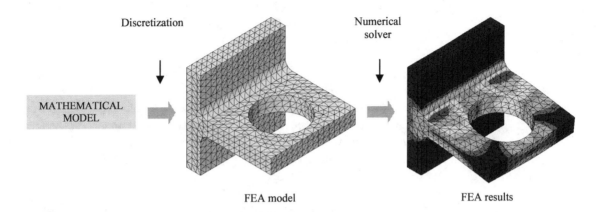

Figure 1-3: Building the finite element model

The mathematical model is discretized into a finite element model. This completes the pre-processing phase. The FEA model is then solved with one of the numerical solvers available in COSMOSWorks.

Solving the finite element model

Having created the finite element model, we now use a solver provided in COSMOSWorks to produce the desired data of interest (Figure 1-3).

Analyzing the results

Often the most difficult step of FEA is analyzing the results. Proper interpretation of results requires that we understand all simplifications (and errors they introduce) in the first three steps: defining the mathematical model, meshing, and solving.

Errors in FEA

The process illustrated in Figure 1-2 and Figure 1-3 introduces unavoidable errors. Formulation of a mathematical model introduces modeling errors (also called idealization errors), discretization of the mathematical model introduces discretization errors, and solving introduces solution errors. Of these three types of errors, only discretization errors are specific to FEA. Modeling errors affecting the mathematical model are introduced before FEA is utilized and can only be controlled by using correct modeling techniques. Solution errors are caused by the accumulation of round-off errors.

A closer look at finite elements

Meshing splits continuous mathematical models into finite elements. The type of elements created by this process depends on the type of geometry meshed. COSMOSWorks offers three types of elements: solid elements for meshing solid geometry, shell elements for meshing surface geometry, and beam elements for meshing wire frame geometry.

Before proceeding we need to clarify an important terminology issue. In CAD terminology, "solid" denotes the type of geometry: solid geometry (as opposed to surface or wire frame geometry). In FEA terminology, it denotes the type of element used to mesh that solid CAD geometry.

Solid elements

The type of geometry that is most often used for analysis with COSMOSWorks is solid CAD geometry. Meshing of this geometry is accomplished with tetrahedral solid elements, commonly called "tets" in FEA jargon. The tetrahedral solid elements in COSMOSWorks can either be first order elements (draft quality), or second order elements (high quality). The user decides whether to use draft quality or high quality elements for meshing. However, as we will soon prove, only high quality elements should be used for an analysis of any importance. The difference between first and second order tetrahedral elements is illustrated in Figure 1-4.

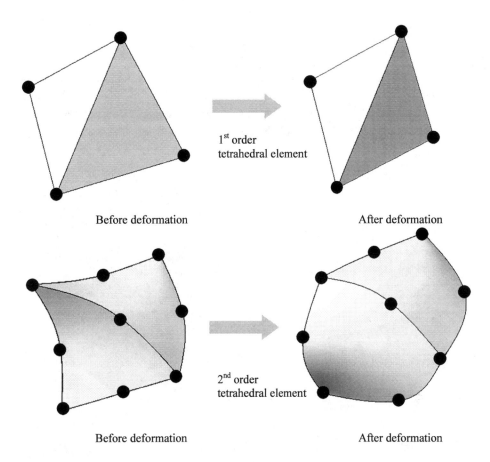

Figure 1-4: Differences between first and second order tetrahedral elements

First and second order tetrahedral elements are shown before and after deformation. Note that the first order element only has corner nodes, while the second order element has both corner and mid-side nodes (one mid-side node is not visible for the second order element in this illustration). Single elements seldom experience deformations of this magnitude, which are exaggerated in this illustration.

In a first order element, edges are straight and faces are flat and must remain this way after deformation. The edges of a second order element before deformation may either be straight or curvilinear, depending on how the element has been mapped to model the actual geometry. Consequently, the faces of a second order element before deformation may be flat or curved.

After deformation, edges of a second order element may either assume a different curvilinear shape or acquire curvilinear shape if they were initially straight. Consequently, faces of a second order element after deformation can be either flat or curved.

First order tetrahedral elements model the linear field of displacement inside their volume, on faces, and along edges. The linear (or first order) displacement field gives these elements their name: first order elements.

If you recall from the Mechanics of Materials, strain is the first derivative of displacement. Since the displacement field is linear, the strain field being the derivative of displacement is constant. Consequently stress field is also constant in first order tetrahedral elements. This situation imposes a very severe limitation on the capability of a mesh constructed with first order elements to model stress distribution of any real complexity. To make matters worse, straight edges and flat faces can not map properly to curvilinear geometry, as illustrated in Figure 1-5.

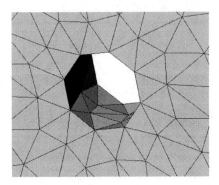

Figure 1-5: Failure of straight edges and flat faces to map to curvilinear geometry

A detail of a mesh created with first order tetrahedral elements. Notice the imprecise element mapping of the hole; flat faces approximate the face of the curvilinear geometry.

Second order tetrahedral elements have ten nodes and model the second order (parabolic) displacement field and first order (linear) stress field in their volume, on faces and along edges. The edges and faces of second order tetrahedral elements can be curvilinear before and after deformation, therefore these elements can map precisely to curved surfaces, as illustrated in Figure 1-6. Even though these elements are more computationally demanding than first order elements, second order tetrahedral elements are used for the majority of analyses with COSMOSWorks.

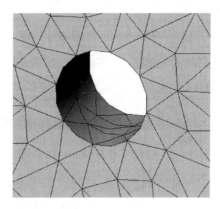

Figure 1-6: Mapping curved surfaces

A detail of a mesh created with second order tetrahedral elements. Second order elements map well to curvilinear geometry.

Shell elements

Shell elements are created by meshing surfaces or faces of solid geometry. Shell elements are primarily used for analyzing thin-walled structures. Since surface geometry does not carry information about thickness, the user must provide this information. Similar to solid elements, shell elements also come in draft and high quality with analogous consequences with respect to their ability to map to curvilinear geometry, as shown in Figure 1-7.

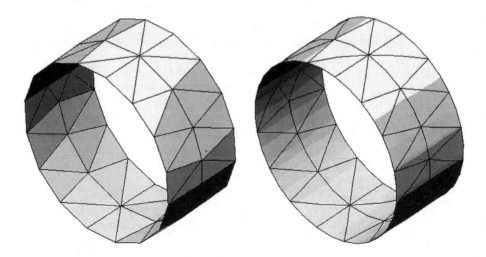

Figure 1-7: First order shell elements (left) and second order shell elements (right)

This shell element mesh on the left was created with first order elements. Notice the imprecise mapping of the mesh to curvilinear geometry. Shell element mesh on the right was created with second order elements, which map correctly to curvilinear geometry.

As in the case of solid elements, first order shell elements model the linear displacements and constant strain and stress. Second order shell elements model the second order (parabolic) displacement and linear strain and stress.

The assumptions of modeling first or second order displacements in shell elements apply only to in-plane directions. The distribution of in-plane stresses across the thickness is assumed to be linear in both first and second order shell elements.

Certain classes of shapes can be modeled using either solid or shell elements, such as the plate shown in Figure 1-8. Often the nature of the geometry dictates what type of element should be used for meshing. For example, a part produced by casting would be meshed with solid elements, while a sheet metal structure would be best meshed with shell elements.

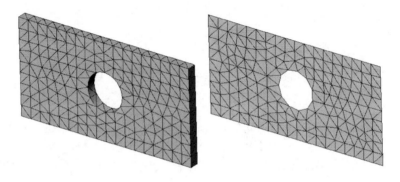

Figure 1-8: Plate modeled with solid elements (left) and shell elements

The actual choice between solids and shells depends on the particular requirements of analysis and sometimes on personal preferences.

Beam elements

Beam elements are created by meshing wire frame geometry. They are the natural choice for meshing weldments. Assumptions about stress distribution in two directions of the beam cross section are made.

A beam element does not have any physical dimensions in the directions normal to its length. It is possible to think of a beam element as a line with assigned beam cross section properties (Figure 1-9).

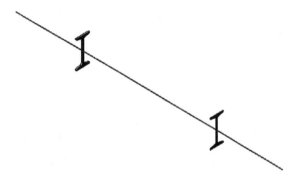

Figure 1-9: Conceptual representation of a beam element

A beam element is a line with assigned properties of a beam cross section as required by beam theory. This illustration conceptualizes how the wire frame defines the cross section of an I-beam and does not represent actual stored geometry.

Figure 1-10 presents the basic library of elements in COSMOSWorks. Solid elements are tetrahedral, shell elements are triangles, and beam elements are straight lines. Elements such as hexahedral solids, quadrilateral shell or curvilinear beams are not available in COSMOSWorks.

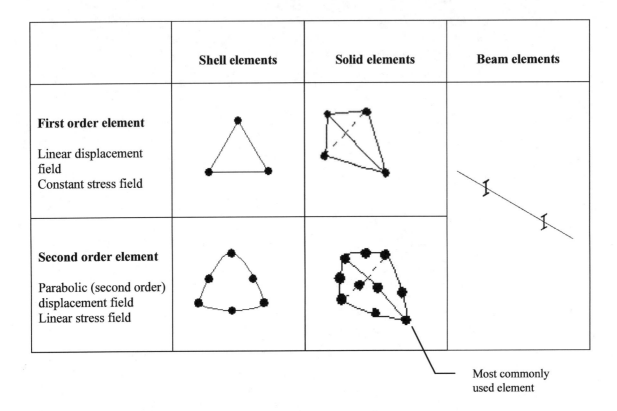

Figure 1-10: COSMOSWorks element library

The vast majority of analyses use the second order tetrahedral element.

The degrees of freedom (DOF) of a node in a finite element mesh (as marked in Figure 1-10) define the ability of the node to perform translation and rotation. The number of degrees of freedom that a node possesses depends on element type. In COSMOSWorks, nodes of solid elements have three degrees of freedom, while nodes of shell elements have six degrees of freedom.

In order to describe transformation of a solid element from the original to the deformed shape, we only need to know three translational components of nodal displacement. In the case of shell elements, we need to know the translational components of nodal displacements and the rotational displacement components.

What is calculated in FEA?

Each degree of freedom of a node in a finite element mesh constitutes an unknown. In structural analysis, nodal degrees of freedom represent displacement components, while in thermal analysis they represent temperatures. Nodal displacements and nodal temperatures are the primary unknowns for structural analysis and thermal analysis, respectively.

Structural analysis finds displacements, strains and stresses. If solid elements are used, then three displacement components (or 3 degrees of freedom) per node must be calculated. With shell elements, six displacement components (or 6 degrees of freedom) must be calculated. Strains and stresses are calculated based on the nodal displacement results.

Thermal analysis finds temperatures, temperature gradients, and heat flow. Since temperature is a scalar value (unlike displacements, which are vectors), then regardless of what type of element is used, there is only one unknown (temperature) to be found for each node. All other thermal results, such as temperature gradient and heat flux, are calculated based on temperature results. The fact that there is only one unknown to be found for each node, rather than three or six, makes thermal analysis less computationally intensive than structural analysis.

How to interpret FEA results

Results of structural FEA are provided in the form of displacements and stresses. But how do we decide if a design "passes" or "fails"? What constitutes a failure?

To answer these questions, we need to establish some criteria to interpret FEA results, which may include maximum acceptable displacements, maximum stress, or the lowest acceptable natural frequency.

While displacement and frequency criteria are quite obvious and easy to establish, stress criteria are not. Let's assume that we need to conduct a stress analysis in order to ensure that stresses are within an acceptable range. To judge stress results, we need to understand the mechanism of potential failure. If a part breaks, what stress measure best describes that failure? COSMOSWorks can present stress results in any desired form. It is up to us to decide which stress measures should be used to analyze results.

Discussion of various failure criteria would be out of the scope of this book. Any textbook on the Mechanics of Materials provides information on this topic. Here we will limit our discussion to three commonly used failure criteria: Von Mises Stress failure criterion, Maximum Shear Stress failure criterion, and Maximum Normal Stress failure criterion.

Von Mises Stress failure criterion

Von Mises stress, also known as Huber stress, is a stress measure that accounts for all six stress components of a general 3-D state of stress (Figure 1-11).

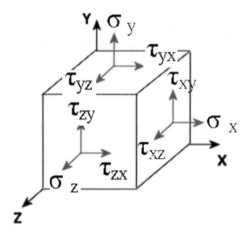

Figure 1-11: *General state of stress represented by three normal stresses: σ_x, σ_y, σ_z and six shear stresses*

Two components of shear stress and one component of normal stress act on each side of this elementary cube. Due to symmetry of shear stresses, the general 3-D state of stress is characterized by six stress components: σ_x, σ_y, σ_z and $\tau_{xy} = \tau_{yx}$, $\tau_{yz} = \tau_{zy}$, $\tau_{xz} = \tau_{zx}$

Von Mises stress σ_{vm}, can be expressed either by six stress components as:

$$\sigma_{vm} = \sqrt{0.5*[(\sigma_x - \sigma_y)^2 + (\sigma_y - \sigma_z)^2 + (\sigma_z - \sigma_x)^2] + 3*(\tau_{xy}^2 + \tau_{yz}^2 + \tau_{zx}^2)}$$

or by three principal stresses (see the paragraph on Maximum Normal Stress failure criterion) as:

$$\sigma_{vm} = \sqrt{0.5*[(\sigma_1 - \sigma_2)^2 + (\sigma_2 - \sigma_3)^2 + (\sigma_3 - \sigma_1)^2]}$$

Note that von Mises stress is a non-negative, scalar value. Von Mises stress is commonly used to present results because the structural safety for many engineering materials showing elasto-plastic properties (for example, steel) can be evaluated using von Mises stress.

The maximum von Mises stress criterion is based on the von Mises-Hencky theory, also known as the Shear-energy theory or the Maximum distortion energy theory. The theory states that a ductile material starts to yield at a location when the von Mises stress becomes equal to the stress limit. In most cases, the yield strength is used as the stress limit. According to von Mises failure criterion the factor of safety FOS is expressed as:

$$FOS = \sigma_{limit} / \sigma_{vm}$$

Maximum Shear Stress failure criterion

Also known as Tresca yield criterion, the maximum shear stress failure criterion is based on the Maximum Shear stress theory. This theory predicts failure of a material to occur when the absolute maximum shear stress (τ_{max}) reaches the stress that causes the material to yield in a simple tension test. The Maximum shear stress criterion is used for ductile materials. According to Maximum Shear stress failure criterion the factor of safety FOS is expressed as:

$$FOS = \sigma_{limit} / (2\tau_{max})$$

Maximum Normal Stress failure criterion

By properly adjusting the angular orientation of the stress cube in Figure 1-11, shear stresses disappear and the state of stress is represented only by three principal stresses: σ_1, σ_2, σ_3, as shown in Figure 1-12. In COSMOSWorks, principal stresses are denoted as P1, P2, and P3.

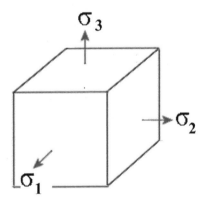

Figure 1-12: General state of stress represented by three principal stresses: σ_1, σ_2, σ_3

This criterion is used for <u>brittle</u> materials. It assumes that the ultimate strength of the material in tension and compression is the same. This assumption is not valid in all cases. For example, cracks considerably decrease the strength of the material in tension while their effect is not significant in compression because the cracks tend to close. Brittle materials do not have a specific yield point and hence it is not recommended to use the yield strength to define the limit stress for this criterion.

This theory predicts failure will occur when σ_1 exceeds stress limit, usually the Ultimate strength. According to maximum principle stress failure criterion the factor of safety FOS is expressed as:

$$FOS = \sigma_{limit} / \sigma_1$$

Units of measure

Internally, COSMOSWorks uses the International System of Units (SI). However, for the user's convenience, the unit manager allows data entry in any of three systems of units: SI, Metric, and English. Results can be displayed using any of the three systems. Figure 1-13 summarizes the available systems of units.

	International System (SI)	Metric (MKS)	English (IPS)
Mass	kg	kg	lb.
Length	m	cm	in.
Time	s	s	s
Force	N	Kgf	lb.
Mass density	kg/m^3	kg/cm^3	$lb./in.^3$
Temperature	K	°C	°F

Figure 1-13: Unit systems available in COSMOSWorks

SI, Metric, and English systems of units can be interchanged when entering data or analyzing results in COSMOSWorks.

Experience indicates that units of mass density are often confused with units of specific gravity. The distinction between these two is quite clear in SI units: Mass density is expressed in $[kg/m^3]$, while specific gravity in $[N/m^3]$. However, in the English system, both specific mass and specific gravity are expressed in $[lb/in^3]$, where $[lb]$ denotes either pound mass or pound force.

As COSMOSWorks users, we are spared much confusion and trouble with systems of units. However, we may be asked to prepare data or interpret the results of other FEA software where we do not have the convenience of the unit manager. Therefore, we will make some general comments about the use of different systems of units in the preparation of input data for FEA models. We can use any consistent system of units for FEA models, but in practice, the choice of the system of units is dictated by what units are used in the CAD model. The system of units in CAD models is not always consistent; length can be expressed in $[mm]$, while mass density can be expressed in $[kg/m^3]$. Contrary to CAD models, in FEA all units *must* be consistent. Inconsistencies are easy to overlook, especially when defining mass and mass density, leading to serious errors.

In the SI system, which is based on meters [m] for length, kilograms [kg] for mass, and seconds [s] for time, all other units are easily derived from these basic units. In mechanical engineering, length is commonly expressed in millimeters [mm], force in Newtons [N], and time in seconds [s]. All other units must then be derived from these basic units: [mm], [N], and [s]. Consequently, the unit of mass is defined as a mass which, when subjected to a unit force equal to 1N, will accelerate with a unit acceleration of 1 mm/s^2. Therefore, the unit of mass in a system using [mm] for length and [N] for force, is equivalent to 1,000 kg or one metric ton. Consequently, mass density is expressed in metric tonnes [$tonne/mm^3$]. This is critically important to remember when defining material properties in FEA software without a unit manager. Notice in Figure 1-14 that an erroneous definition of mass density in [kg/m^3] rather than in [$tonne/mm^3$] results in mass density being one trillion (10^{12}) times higher (Figure 1-14).

System SI	[m], [N], [s]
Unit of mass	kg
Unit of mass density	kg/m^3
Density of aluminum	2794 kg/m^3

System of units derived from SI	[mm], [N], [s]
Unit of mass	tonne
Unit of mass density	tonne/mm^3
Density of aluminum	2.794 x 10^{-9} tonne/mm^3

English system (IPS)	[in], [LB], [s]
Unit of mass	slug/12
Unit of mass density	slug/12/in^3
Density of aluminum	2.614 x 10^{-4} slug/12/in^3

Figure 1-14: Mass density of aluminum in the three systems of units

Comparison of numerical values of mass densities of 1060 aluminum alloy defined in the SI system of units with the system of units derived from SI, and with the English (IPS) system of units.

Using on-line help

COSMOSWorks features very extensive on-line Help and Tutorial functions, which can be accessed from the Help menu in the main COSMOSWorks tool bar (Figure 1-15).

Figure 1-15: Accessing the on-line Help and Tutorial

On-line Help and Tutorial can be accessed from the main SolidWorks toolbar.

Limitations of COSMOSWorks Professional

We need to appreciate some important limitations of COSMOSWorks 2006 Professional: material is assumed as linear, and loads are static. At the same time we should note that COSMOSWorks 2006 Professional is no longer limited to small deformation analysis (see Chapter 17).

Linear material

Whatever material we assign to the analyzed parts or assemblies, the material is assumed as linear, meaning that stress is proportional to strain (Figure 1-16).

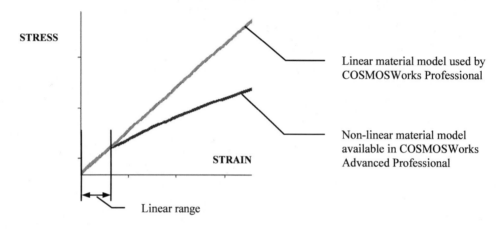

Figure 1-16: Linear material model assumed in COSMOSWorks

In all materials used by COSMOSWorks Professional, stress is linearly proportional to strain. The linear range is where linear and non-linear material models are not significantly different.

Using a linear material model, the maximum stress magnitude is not limited to yield or to ultimate stress as it is in reality.

Material yielding is not modeled, and whether or not yield may in fact be taking place can only be established based on the stress magnitudes reported in results. Most analyzed structures experience stresses below the yield stress, and the factor of safety is most often related to the yield stress. Therefore, the analysis limitations imposed by linear material seldom impede COSMOSWorks Professional users.

Static loads

All structural loads and restraints are assumed not to change with time. Dynamic loading conditions cannot be analyzed with COSMOSWorks Professional (the only exception is **Drop Test** analysis). This limitation implies that loads are applied slowly enough to ignore inertial effects. Dynamic analysis can be performed with COSMOSWorks Advanced Professional.

2: Static analysis of a plate

Topics covered

- Using COSMOSWorks interface
- Linear static analysis with solid elements
- The influence of mesh density on results
- Finding reaction forces
- Controlling discretization errors by the convergence process
- Presenting FEA results in desired format

Project description

A steel plate is supported and loaded, as shown in Figure 2-1. We assume that the support is rigid (this is also called built-in support or fixed support) and that the 100,000 N tensile load is uniformly distributed along the end face, opposite to the supported face.

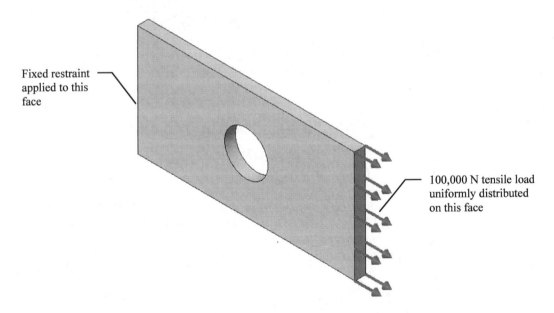

Figure 2-1: SolidWorks model of a rectangular plate with a hole

We will perform displacement and stress analysis using meshes with different element sizes. Note that repetitive analysis with different meshes does *not* represent standard practice in FEA. We will repeat the analysis using different meshes only as a learning tool to gain more insight into how FEA works.

Procedure

In SolidWorks, open the model file called HOLLOW PLATE. Verify that COSMOSWorks is selected in the **Add-Ins** list. To start COSMOSWorks, select the COSMOSWorks Manager tab, as shown in Figure 2-2.

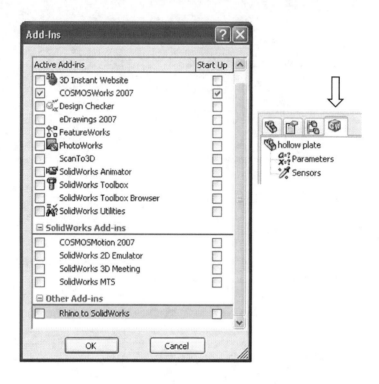

Figure 2-2: Add-Ins list and COSMOSWorks Manager tab

Verify that COSMOSWorks is selected in the list of Add-Ins (left), and then select the COSMOSWorks Manager tab (right).

To create an FEA model, solve it, and analyze the results, we will use a graphical interface. You can also do this by making the appropriate choices in the COSMOSWorks menu. To call up the menu, select COSMOSWorks from the main tool bar of SolidWorks (Figure 2-3).

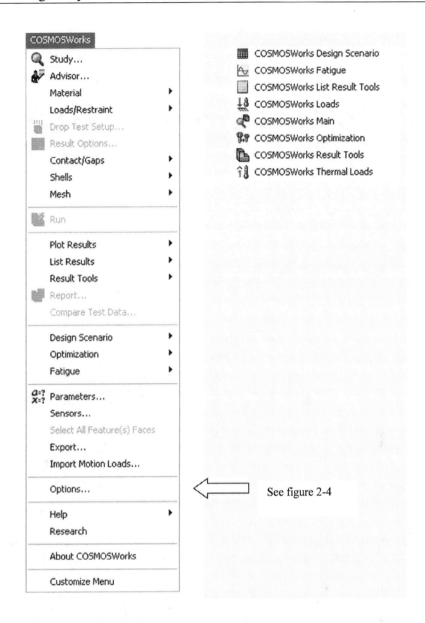

Figure 2-3: COSMOSWorks menu (left) and COSMOSWorks tool bars (right)

All functions used for creating, solving, and analyzing a model can be executed either from the COSMOSWorks menu, the COSMOS tool bars, or the graphical interface in the COSMOSWorks Manager window. In this book we use the graphical interface.

Before we create the FEA model, let's review the **Options** window in COSMOSWorks (Figure 2-4). This window can be accessed from the COSMOSWorks main menu (Figure 2-3).

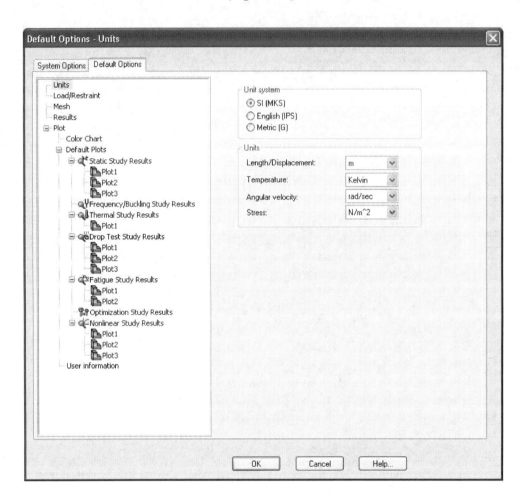

Figure 2-4: COSMOSWorks Options window; shown is Default Options tab

The COSMOSWorks Preferences window has System Options and Default Options tabs. Review both tabs before proceeding with the exercise. Note that Default Plots can be grouped into sub-folders which are created by right-clicking on the Static Study Results folder, Thermal Study Results folder etc.

As shown above, we use the SI system as specified in Default Options tab.

Creation of an FEA model always starts with the definition of a study. To define a study, right-click the *Part* icon in the COSMOSWorks Manager window and select **Study...** from the pop-up menu. In this exercise, the *Part* icon is called *hollow plate*. Figure 2-5 shows the required selections in the study definition window: the analysis type is **Static**, the mesh type is **Solid mesh**. The study name is *tensile load 01*.

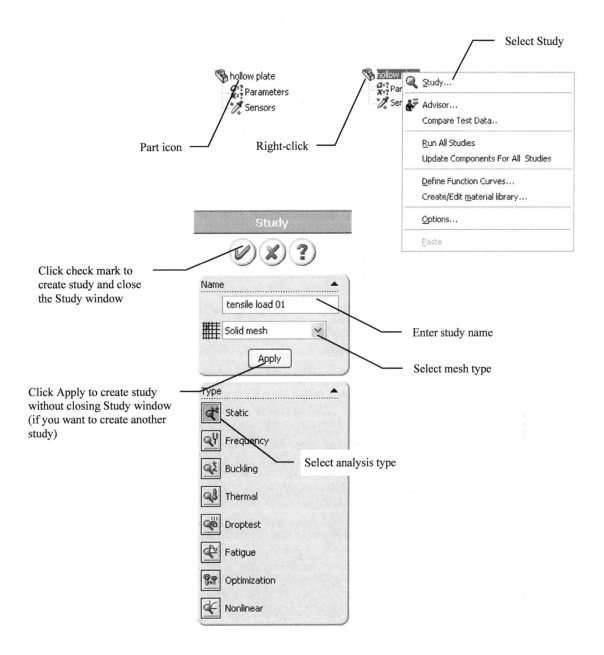

Figure 2-5: Study window

To display the Study window (bottom), right-click the Part icon in the COSMOSWorks Manager window (top left), and then from the pop-up menu (top right), select Study....

When a study is defined, COSMOSWorks automatically creates a study folder (in this case named *tensile load 01*) and places several sub-folders in it (some of these sub-folders are empty because their contents must by defined by the user), as shown in Figure 2-6. Not all folders are used in each analysis. In this exercise, we will use the *Solids* folder to define and assign material properties, the *Load/Restraint* folder to define loads and restraints, and the *Mesh* folder to create the finite element mesh.

Figure 2-6: *Study* folders

COSMOSWorks automatically creates a study folder, called tensile load 01, with the following sub folders: Solids, Load/Restraint, Design Scenario, Contact/Gaps, Mesh, and Report folder. The Design Scenario, Contact/Gaps and Report folders as well as Parameters and Sensors folders, which are automatically created prior to study definition, will not be used will not be used in this exercise.

We are now ready to define the analysis model. This process generally consists of the following steps:

❑ Geometry preparation

❑ Material properties assignment

❑ Restraints application

❑ Load application

In this case, the model geometry does not need any preparation (it is already very simple), so we can start by assigning material properties.

As illustrated in Figure 2-7, you can assign material properties to the model several ways:

❑ Right-clicking the *Solids* folder to apply material to all components in the *Solids* folder.

❑ Right-clicking the *hollow plate* folder, which is located in the *Solids* folder to apply material to all bodies in the *hollow plate* folder.

❑ Right-clicking the *Body 1* icon, which is located in the *hollow plate* folder to apply material to *Body 1* which, in our case, is the only body in *hollow plate* folder.

In this exercise there is only one component which consists of one body (take this opportunity to notice that COSMOSWorks supports analysis of multiple bodies). Therefore, the above three ways of applying material properties produce identical results.

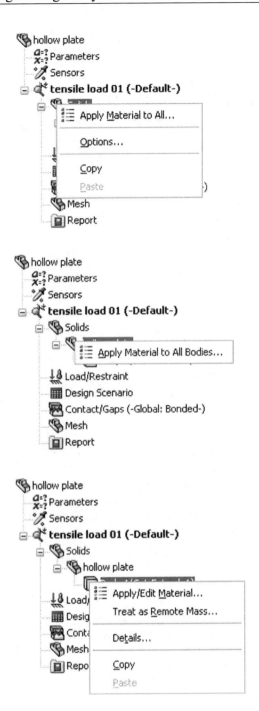

Figure 2-7: *Applying material properties*

Since the model only consists of one component with one body, the material property can be applied in any of the three ways shown here.

Let's use the first method: **Apply Material to All** to open the **Material** window shown in Figure 2-8.

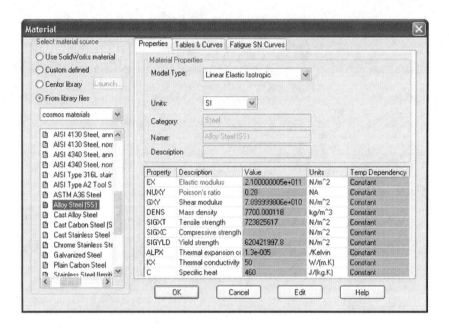

Figure 2-8: Material window

In the Material dialog box, the properties are highlighted to indicate the mandatory and optional properties. A red description (Elastic modulus, Poisson's ratio) indicates the property that is mandatory based on the active study type and the material model. A blue description (Mass density, Tensile strength, Compressive strength, Yield strength, Thermal expansion coefficient) indicates an optional property. A black description (Thermal conductivity, Specific heat) indicates property not applicable to the current study.

Select **From library files** in the **Select material source** area, and then select **Alloy Steel**. Select **SI** units under **Properties** tab (other units could be used as well). Notice that the Solids folder now shows a check mark and the name of the selected material to indicate that a material has successfully been assigned. If needed, you could define your own material by selecting **Custom Defined** material.

Note that material definition consists of two steps:

❑ Material selection (or material definition if custom material is used)

❑ Material assignment to either all solids in the model or to selected bodies of a multi-body part, or to selected components of an assembly

We should also notice that if a material has been defined for a SolidWorks part model, material definition is automatically transferred to the COSMOSWorks model. Assigning a material to the SolidWorks model is actually a preferred modeling technique, especially when working with an assembly consisting of parts with many different materials. We will do this in later exercises.

Having defined the material, we now move to defining the loads and restraints. To display the pop-up menu that lists the options available for defining loads and restraints, right-click the *Load/Restraint* folder in the *tensile load 01* study (Figure 2-9).

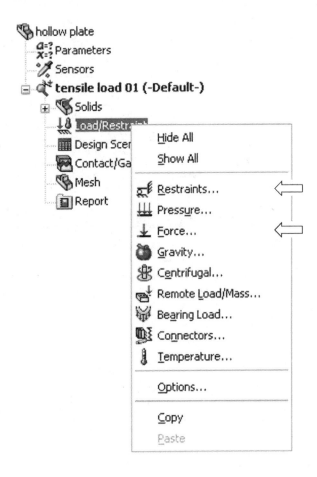

Figure 2-9: Pop-up menu for the *Load/Restraint* folder

The arrows indicate the selections used in this exercise.

To define the restraints, select **Restraints...** from the pop-up menu displayed in Figure 2-9. This action opens the **Restraint** window shown in Figure 2-10.

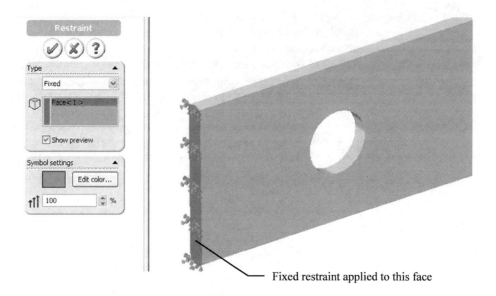

Figure 2-10: Restraint window

In Restraints, select Fixed as Restraint type and also select the model face where the restraint is to be applied. In the graphic window, note the symbols of the applied restraint. See the table on next page for more explanations.

You can rotate the model in order to select the face where the restraints are applied. The Rotate, Pan, Zoom and all other view functions in COSMOSWorks work the same as in SolidWorks.

In the **Restraint** window, select **Fixed** as the type of restraint to apply. The following chart reviews other choices offered in the **Restraint** window.

Restraint Type	Definition
Fixed	Also called built-in or rigid support, all translational and all rotational degrees of freedom are restrained.
Immovable (No translations)	Only translational degrees of freedom are constrained, while rotational degrees of freedom remain unconstrained. If solid elements are used (like in this exercise), **Fixed** and **Immovable** restraints have the same effect because solid elements do not have rotational degrees of freedom.
Symmetry	This option automatically applies symmetry boundary conditions to a flat face. Translation in the direction normal to the face is restrained and rotations about axes aligned with the face are restrained.
Roller/Sliding	This option specifies that a planar face can move freely on its plane but not in the direction normal to its plane. The face can shrink or expand under loading.
Hinge	This option applies only to cylindrical face and specifies that the cylindrical face can only rotate about its own axis. The radius and the length in the axial direction of the cylindrical face remain constant under loading. This condition is similar to selecting the **On cylindrical face** restraint type and setting the radial and axial components to zero.
Use reference geometry	This option restrains a face, edge, or vertex only in certain directions, while leaving the other directions free to move. You can specify the desired directions of restraint in relation to the selected reference plane or reference axis.
On flat face	This option provides restraints in selected directions, which are defined by the three principal directions of the flat face where restraints are being applied.
On cylindrical face	This option is similar to **On flat face**, except that the three principal directions of a cylindrical face define the directions of restraints.
On spherical face	Similar to **On flat face** and **On cylindrical face**. The three principal directions of a spherical face define the directions of applied restraints.
Cyclic symmetry	This option allows analysis of a model with circular patterns around an axis by modeling only a representative segment. The segment can be a part or an assembly. The geometry, restraints, and loading conditions must be similar for all other segments making up the model. Turbines, fans, flywheels, and motor rotors can usually be analyzed using cyclic symmetry.

When a model is fully supported (as it is in our case), we say that the model does not have any rigid body motions (the term "rigid body modes" is also used) meaning it cannot move without experiencing deformation.

Note that the presence of restraints in the model is manifested by both the restraint symbols (showing on the restrained face) and by the automatically created icon, *Restraint-1*, in the *Load/Restraint* folder. The display of the restraint and load symbols can be turned on and off by either:

❑ Using the **Hide All** or **Show All** commands in the pop-up menu shown in Figure 2-9, or

❑ Right-clicking the restraint or load icon individually to display a pop-up menu and then selecting **Hide** or **Show** from the pop-up menu.

Now define the load by selecting **Force** from the pop-up menu shown in Figure 2-9. This action opens the **Force** window (Figure 2-11).

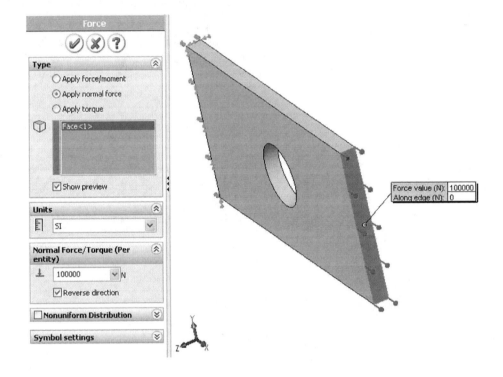

Figure 2-11: Force window

The Force window displays the selected face where tensile force is applied. This illustration also shows symbols of applied restraint and load.

In the Type area, select the **Apply normal force** button in order to load the model with 100,000 N tensile force uniformly distributed over the end face, as shown in Figure 2-11. Check **Reverse direction** option to apply a tensile load.

Generally, forces can be applied to faces, edges, and vertexes using different methods, which are reviewed below:

Force Type	Definition
Apply force/moment	This option applies force or moment to a face, edge, or vertex in the direction defined by selected reference geometry.
	Note that moment can be applied only if shell elements are used. Shell elements have six degrees of freedom per node: three translations and three rotations, and can take a moment load. Solid elements only have three degrees of freedom (translations) per node, and therefore cannot take a moment load directly.
	If you need to apply moment to solid elements, it must be represented with appropriately applied forces.
Apply normal force	Available for flat faces only, this option applies load in the direction normal to the selected face.
Apply torque	Used for cylindrical faces, this option applies torque (expressed by traction forces) about a reference axis using the right-hand rule.

The presence of load(s) is visualized by arrows symbolizing the load and by automatically created icons *Force-1* in the *Load/Restraint* folder.

Try using the click-inside technique to rename *Restraint-1* and *Force-1* icons. Note that renaming using the click-inside technique works on all icons in the COSMOSWorks Manager.

The model is now ready for meshing. Before creating the mesh, let's make a few observations about defining:

- Geometry
- Material properties
- Loads
- Restraints

Geometry preparation is a well-defined step with few uncertainties. Geometry that is simplified for analysis can be checked visually by comparing it with the original CAD model.

Material properties are most often selected from the material library and do not account for local defects, surface conditions, etc. Therefore, definition of material properties usually has more uncertainties than geometry preparation.

The definition of loads is done in a few quick menu selections, but involves many assumptions. Factors such as load magnitude and distribution are often only approximately known and must be assumed. Therefore, significant idealization errors can be made when defining loads.

Defining restraints is where severe errors are most often made. For example, it is easy enough to apply a fixed restraint without giving too much thought to the fact that a fixed restraint means a rigid support – a mathematical abstraction. A common error is over-constraining the model, which results in an overly stiff structure that underestimates displacements and stresses. The relative level of uncertainties in defining geometry, material, loads, and restraints is qualitatively shown in Figure 2-12.

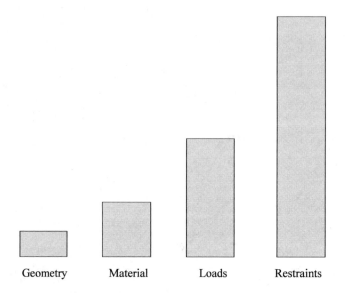

Figure 2-12: Qualitative comparison of uncertainty in defining geometry, material properties, loads, and restraints

The level of uncertainty (or the risk or error) has no relation to time required for each step, so the message in Figure 2-12 may be counterintuitive. In fact, preparing CAD geometry for FEA may take hours, while applying restraints takes only a few mouse clicks.

In all of the examples presented in this book, we assume that definitions of material properties, loads, and restraints represent an acceptable idealization of real conditions. However, we need to point out that it is the responsibility of FEA user to determine if all those idealized assumptions made during the creation of the mathematical model are indeed acceptable.

We are now ready to mesh the model, but first verify under **Default Options**, **Mesh** tab (Figure 2-13) that **High** mesh quality is selected. The **Options** window can be opened from COSMOSWorks main menu, as shown in Figure 2.13.

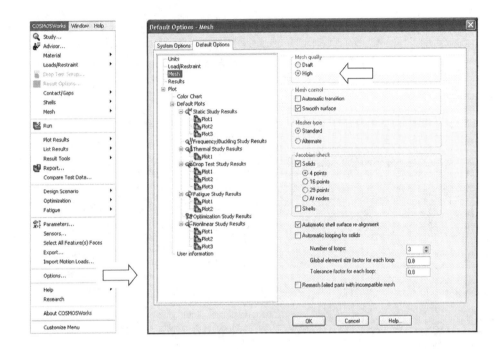

Figure 2-13: Mesh tab in the Options window

Use this window to verify that mesh quality is set to High

The difference between **High** and **Draft** mesh quality is that:

- Draft quality mesh uses first order elements
- High quality mesh uses second order elements

Differences between first and second order elements were discussed in Chapter 1.

Now, right-click the **Mesh** folder to display the pop-up menu (Figure 2-14).

Figure 2-14: Mesh pop-up menu

In the pop-up menu, select **Create...** to open the **Mesh** window. This window offers a choice of element size and element size tolerance.

In this exercise, we will study the impact of mesh size on results. Therefore, we will solve the same problem using three different meshes: coarse, medium (default) and fine. Figure 2-15 shows the respective selection of meshing parameters to create the three meshes.

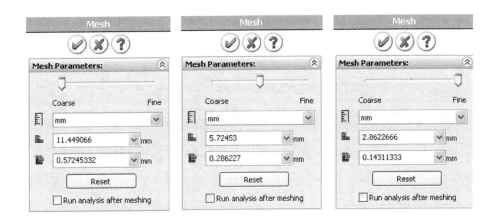

Figure 2-15: Three choices for mesh density from left to right: coarse, medium (default), and fine

The medium mesh density, shown in the middle window in Figure 2-15, is the default that COSMOSWorks proposes for meshing our model. The element size of 5.72 mm and the element size tolerance of 0.286mm are established automatically based on the geometric features of the SolidWorks model. The 5.72-mm size is the characteristic element size in the mesh, as explained in Figure 2-16. The element size tolerance is the allowable variance of the actual element sizes in the mesh.

Mesh density has a direct impact on the accuracy of results. The smaller the elements, the lower the discretization errors, but the meshing and solving time both take longer. In the majority of analyses with COSMOSWorks, the default mesh settings produce meshes that provide acceptable discretization errors, while keeping solution times reasonably short.

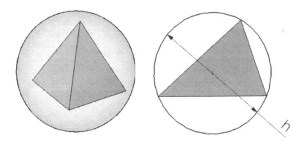

Figure 2-16: Characteristic element size for a tetrahedral element

The characteristic element size of a tetrahedral element is the diameter h of a circumscribed sphere (left). This is easier to illustrate with the 2-D analogy of a circle circumscribed on a triangle (right).

Right-click the Mesh icon again and select **Create...** to open the **Mesh** window.

With the **Mesh** window open, set the slider all the way to the left (as illustrated in Figure 2-15, left) to create a coarse mesh, and click the green check mark button. The mesh will be displayed as shown in Figure 2-17.

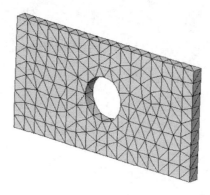

Figure 2-17: Coarse mesh created with second order, solid tetrahedral elements

You can control the mesh visibility by selecting Hide Mesh or Show Mesh from the pop-up menu shown in Figure 2-14.

To start the solution, right-click the *tensile load 01* study folder to display a pop-up menu (Figure 2-18). Select **Run** to start the solution.

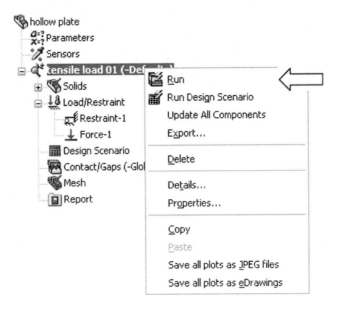

Figure 2-18: Pop-up menu for the *Study* folder

Start the solution by right-clicking the tensile load 01 icon to display a pop-up menu. Select Run to start the solution.

The solution can be executed with different properties, which will be investigated in later chapters. You can monitor the solution progress while the solution is running (Figure 2-19).

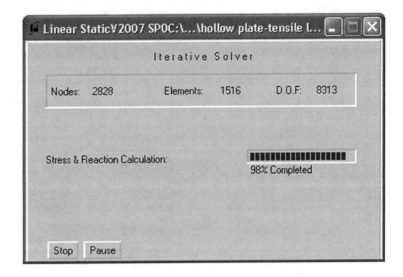

Figure 2-19: Solution Progress window

The Solver reports solution progress while the solution is running

If the solution fails, the failure is reported as shown in Figure 2-20.

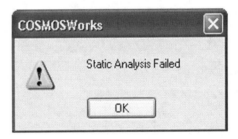

Figure 2-20: Failed solution warning window

When the solution completes, COSMOSWorks automatically creates three plots in the *Results* folder:

- *Stress1*   showing von Mises stresses
- *Displacement1* showing resultant displacements
- *Strain1*   showing equivalent strain

as shown in Figure 2-21.

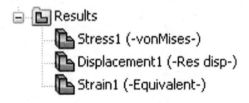

Figure 2-21: Automatically created *Results* folders

If desired, you can add more plots to the *Results* folder. You can also create subfolders in the *Results* folder to group plots (Figure 2-22).

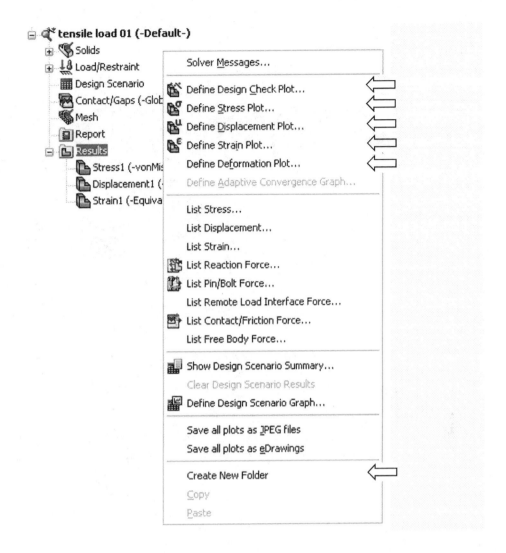

Figure 2-22: More plots and folders can be added to *Results* folder

Right-clicking on Results activates this pop up menu from where plots and folders may be added to the Results folder.

To display stress results, double-click on the *Stress1* icon in the *Results* folder or right-click it and select **Show** from the pop-up menu. The default stress plot is shown in Figure 2-23.

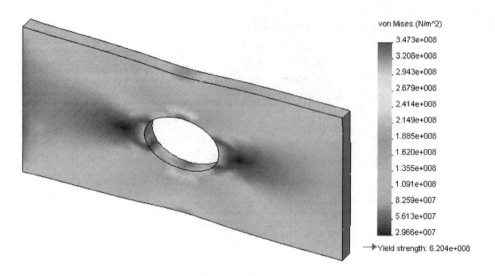

Figure 2-23: Stress plot displayed using default stress plot settings

Von Mises stress results are shown by default in the stress plot window. Notice that results are shown in [Pa] and the highest stress 347.3×10^8 Pa is below the material yield strength 620.4×10^8 Pa. The actual numerical results may differ slightly depending on solver, software version and service pack used.

Once the stress plot is showing, right-click the stress plot icon to display the pop-up menu featuring different plot display options (Figure 2-24).

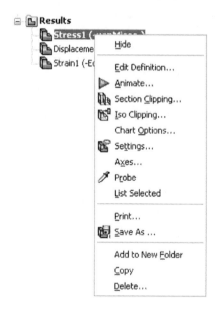

Figure 2-24: Pop-up menu with plot display options

Any plot can be modified using selections from this pop-up menu.

We will now examine how to modify stress plot using **Edit Definition**, **Chart Options** and **Settings,** which are all selectable from the pop-up menu in Figure 2-24.

Select **Edit Definition** to open the **Stress Plot** window, change the units to MPa then close the window. In the **Chart Options** window, change the range of the displayed stress results from **Automatic** to **Defined**, and define the range as from 0 to 300 MPa. In **Settings,** select **Discrete** in **Fringe options** and **Mesh** in **Boundary options**. All these selections are shown in Figure 2-25.

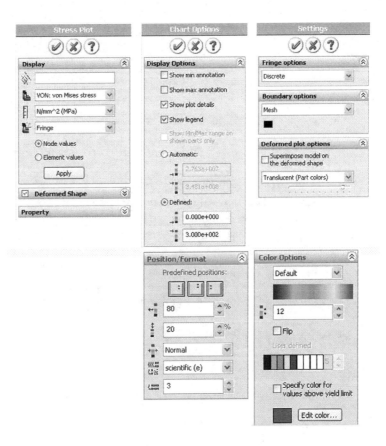

Figure 2-25: **Stress Plot** (invoked by **Edit Definition...**), **Chart Options** and **Settings** windows we use in this exercise to modify stress plot display

Color Options section belongs to Chart Options window.

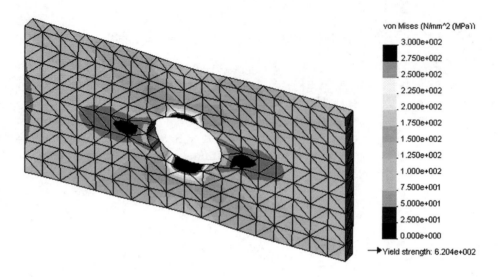

Figure 2-26: Modified stress plot results

The modified stress plot using selections shown in windows in Figure 2-25.

The Stress plot in Figure 2-26 shows node values, also called averaged stresses. Element values (or non-averaged stresses) can be displayed by proper selection in the **Stress Plot** window. Node values are most often used to present stress results. See Chapter 3 and the glossary of terms in Chapter 22 for more comments on node and element values of stress results.

Before you proceed, investigate other selections available in the windows shown in Figure 2-24 and the pop-up menu shown in Figure 2-23.

We will now review the displacement, strain, and deformation results. All of these plots are created and modified in the same way. Sample results are shown in:

❑ Figure 2-27 (displacement)
❑ Figure 2-28 (strain)
❑ Figure 2-29 (deformation)

Before proceeding, create a deformation plot using the pop up menu in Figure 2-22.

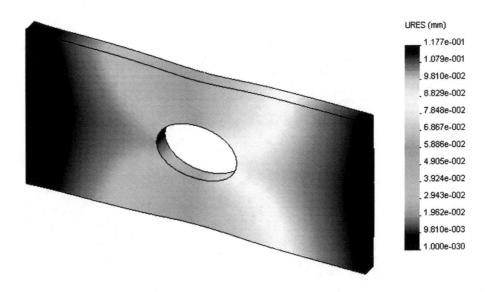

Figure 2-27: Displacement results using Continuous fringe options

Also try the Discrete fringe option selectable in Settings from plot pop up menu.

The plot in Figure 2-27 shows the deformed shape in an exaggerated scale. You can change the display from undeformed to deformed and modify the scale of deformation in the **Displacement Plot** window activated by right-clicking plot icon, then select **Edit Definition**.

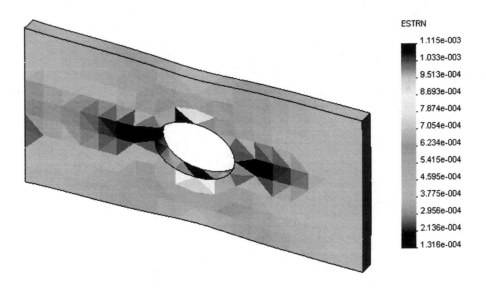

Figure 2-28: Strain results

Strain results are shown here using Element values.

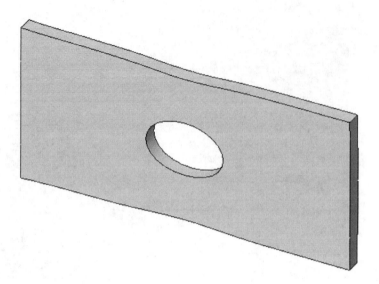

Figure 2-29: Deformation results

Note that all previous results were also showing the model in a deformed shap,e because the deformed shape display was selected in plot definition.

Now, define a **Design Check Plot** using the menu shown in Figure 2-22. Definition of Design Check Plot proceeds in three steps. Follow step 1, step 2, and step 3, as shown in Figure 2-30.

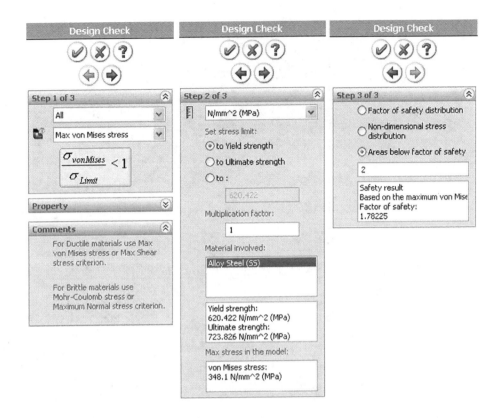

Figure 2-30: Three windows show three steps in the Design Check plot definition

Step one selects the failure criterion, step 2 selects display units and sets the stress limit, step three selects what will be displayed in the plot (here we select areas below the factor of safety 2).

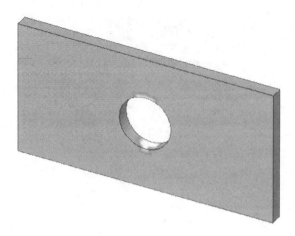

Figure 2-31: Red color (shown as white in this grayscale illustration) displays the areas where the factor of safety falls below 2

We have completed the analysis with a coarse mesh and now wish to see how a change in mesh density will affect the results. Therefore, we will repeat the analysis two more times using medium and fine density meshes respectively. We will use the settings shown in Figure 2-15. All three meshes used in this exercise (coarse, medium, and fine) are shown in Figure 2-32.

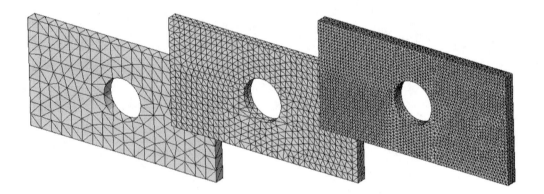

Figure 2-32: Coarse, medium, and fine meshes

Three meshes used to study the effects of mesh density on results.

To compare the results produced by different meshes, we need more information than is available in the plots. Along with the maximum displacement and the maximum von Mises stress, for each study we need to know:

❑ The number of nodes in the mesh.

❑ The number of elements in the mesh.

❑ The number of degrees of freedom in the model.

The information on the number of nodes and number of elements can be found in **Mesh Details** (Figure 2-33).

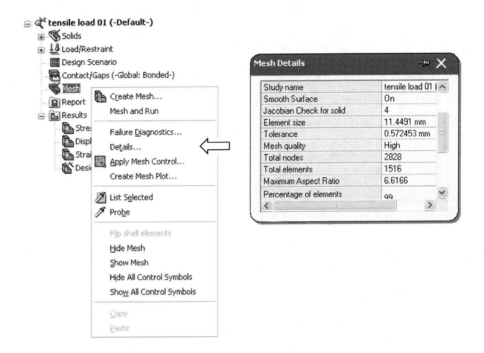

Figure 2-33: Meshing details window

Right-click on Mesh, Details… to display the Mesh Details window.

The number of degrees of freedom is shown in the COSMOS data base. Look for a file with the extension OUT, which is located in the **Results** folder specified in the **Options** window under **Default Options** (Figure 2-34).

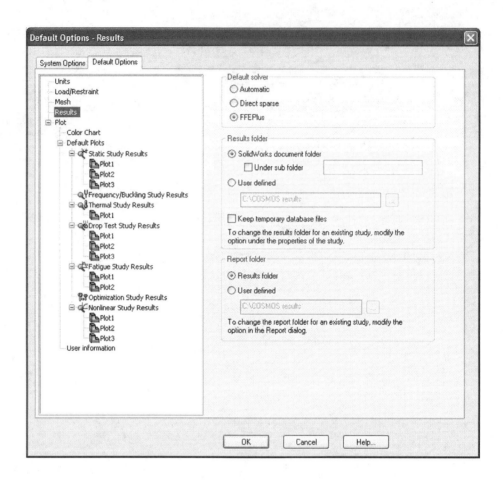

Figure 2-34: COSMOSWorks data base is located in Work directory specified under Results Options in Options window

Here, the SolidWorks document folder is used to store analysis results.

Using Windows Explorer, find the file named hollow plate-tensile load 01.OUT. Open it with a text editor (e.g. Notepad) and find the number of degrees of freedom (referred to in the file as the number of equations).

Note that the OUT file is available only while the model is opened in COSMOSWorks. Upon exiting from COSMOSWorks (which is done by means of deselecting COSMOSWorks from the list of add-ins, or by closing the SolidWorks model), all data base files are compressed into one file with the extension CWR. In our case, the file is named hollow plate-tensile load 01.CWR. If more than one study has been executed, then there is one CWR file for each study. Compressing the entire data base into one file allows for convenient backup of COSMOSWorks results.

Yet another and most convenient way to find the number of nodes, elements, and degrees of freedom is to use the pop up menu activated by right-clicking on the *Results* folder (Figure 2-22). Right-click on **Solver Messages** to open the **Solver Messages** window showing, among other information, the number of nodes, elements and degrees of freedom (Figure 2-35).

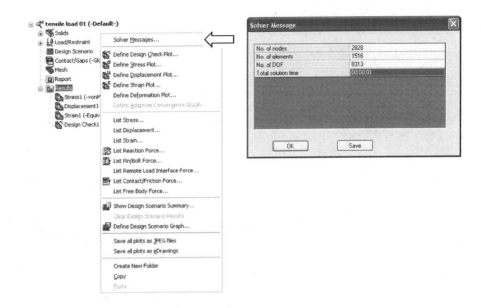

Figure 2-35: Solver Message window lists information pertaining to the solved study.

Now create and run two more studies: *tensile load 02* with medium (default) element size and *tensile load 03* with fine element size, as shown in Figure 2-15.

To create a new study we could just repeat the same steps as before but an easier way is to copy a study. To copy a study, right-click the study folder, then select **Copy**. Next, right-click the model icon (hollow plate) and select **Paste** to invoke **Define Study Name** window, where the name of a new study is defined in (Figure 2-36).

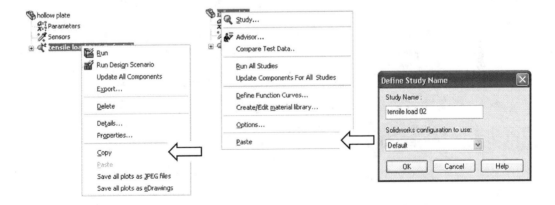

Figure 2-36: A study can be copied into another study in three steps as shown

An alternative way to copy a study is to use Copy (Ctrl-C) and Paste (Ctrl-V) technique of the Windows operating system. Yet another way is to drop study folder *tensile load 01* into the *hollow plate* icon (top of COSMOS Manager).

Note that all definitions in a study (material, restraints, loads, mesh) can also be copied individually from one study to another using one of the above methods.

A study is copied complete with results and plot definitions. Before remeshing study *tensile load 02* with the default element size mesh, you must acknowledge the warning message shown in Figure 2-37.

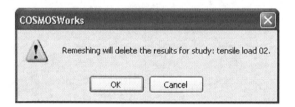

Figure 2-37: Remeshing deletes any existing results in the study

The summary of results produced by the three models is shown in Figure 2-38.

Mesh density	Max. displacement magnitude [mm]	Max. von Mises stress [MPa]	Number of degrees of freedom	Number of elements	Number of nodes
Coarse	0.1178	348	8313	1516	2828
Medium	0.1180	378	36285	7082	12228
Fine	0.1181	376	250473	54889	83986

Figure 2-38: Summary of results produced by the three meshes

Note that these results are based on the same problem. Differences in the results arise from the different mesh densities used in studies tensile load 01, tensile load 02, and tensile load 03.

Figures 2-39 and 2-40 show the maximum displacement and the maximum von Mises stress as functions of the number of degrees of freedom. The number of degrees of freedom is in turn a function of mesh density.

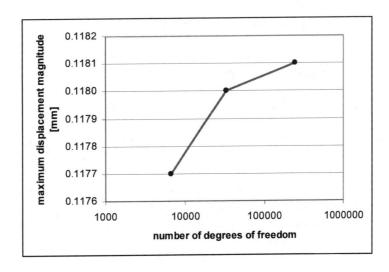

Figure 2-39: Maximum displacement magnitude

Maximum displacement magnitude is plotted as a function of the number of degrees of freedom in the model. The three points on the curve correspond to the three models solved. Straight lines connect the three points only to visually enhance the graph.

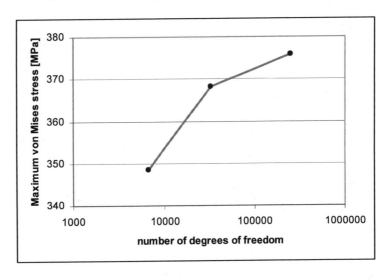

Figure 2-40: Maximum von Mises stress

Maximum von Mises stress is plotted as a function of the number of degrees of freedom in the model. The three points on the curve correspond to the three models solved. Straight lines connect the three points only to visually enhance the graph.

Having noticed that the maximum displacement increases with mesh refinement, we can conclude that the model becomes "softer" when smaller elements are used. With mesh refinement, a larger number of elements allows for better approximation of the real displacement and stress field. Therefore, we can say that the artificial constraints imposed by element definition become less imposing with mesh refinement.

Displacements are always the primary unknowns in structural FEA, and stresses are calculated based on displacement results. Therefore, stresses also increase with mesh refinement. If we continued with mesh refinement, we would see that both the displacement and stress results converge to a finite value which is the solution of the mathematical model. Differences between the solution of the FEA model and the mathematical model are due to discretization errors, which diminish with mesh refinement.

We will now repeat our analysis of the hollow plate by using prescribed displacements in place of a load. Rather than loading it with a 100,000 N force that has caused a 0.118 mm displacement of the loaded face, we will apply a prescribed displacement of 0.118 mm to this face to see what stresses this causes. For this exercise, we will use only one mesh with default (medium) mesh density.

Define the fourth study, called ***prescribed displ***. The easiest way to do this is to copy the definitions from the ***tensile load 02*** study. The definition of material properties, the fixed restraint to the left-side end-face and mesh are all identical to the previous design study. We need to delete the load (right-click load icon and select **Delete**) and apply in its place prescribed displacement.

To apply the prescribed displacement to the right-side end-face, open the **Restraint** window, select the face and define displacement as shown in Figure 2-38. Check *Reverse direction* to obtain displacement in the tensile direction. Note that direction of prescribed displacement is indicated by a restraint symbol. The size and color of this symbol can be changed using *Symbol Settings* (Figure 2-41). The color and size of all load and restraint symbols is controlled the same way.

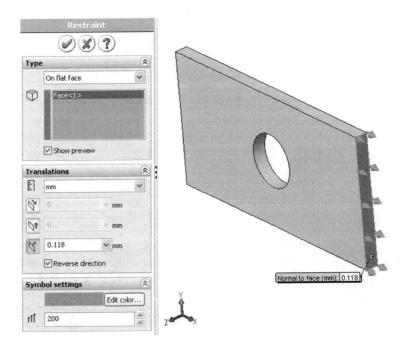

Figure 2-41: Restraint definition window

The prescribed displacement of 0.118 mm is applied to the same face where the tensile load of 100,000 N had been applied.

Once prescribed displacement is defined to the end face, it overrides any previously applied loads to the same end face. While it is better to delete the load in order to keep the model clean, the load has no effect if prescribed displacement is applied to the same entity and in the same direction.

Figures 2-39 and 2-40 compare displacement and stress results for model loaded with force and with prescribed displacement.

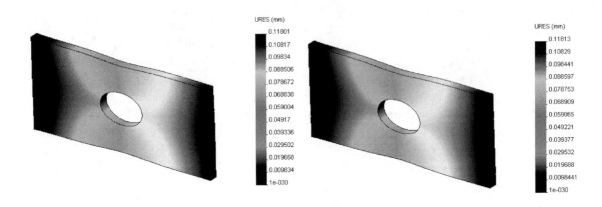

Figure 2-39: Comparison of displacement results

Displacement results in the model with the force load are displayed on the left, and displacement results in the model with the prescribed displacement load are displayed on the right.

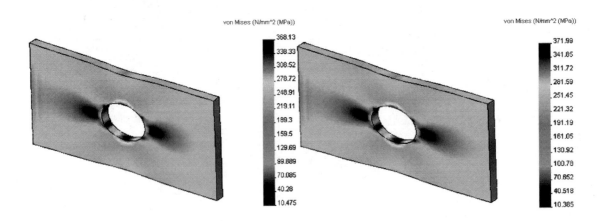

Figure 2-40: Von Mises stress results

Von Mises stress results with load applied as force are displayed on the left and Von Mises stress results with load applied as prescribed displacement are displayed on the right.

Note the numerical format of results. You can change the format in the **Chart Options** window (Figure 2-25).

Results produced by applying a force load and by applying a prescribed displacement load are very similar, but not identical. The reason for this discrepancy is that in the force load model, the loaded face is allowed to deform. In the prescribed displacement model, this face remains flat, even though it experiences displacement as a whole. Also, while the prescribed displacement of 0.118 mm applies to the entire face in the prescribed displacement model, it is only seen as a maximum displacement in one point in the force load model.

We conclude our analysis of the hollow plate by examining the reaction forces. Let's use results of study *tensile load 02* for this analysis. Open the study *tensile load 02* folder and right-click *Results*. From the pop-up menu, select *List Reaction Force* to open **Result Force** window. Select face where fixed restraint is applied and click the **Update** button. Information on reaction forces will be displayed as shown in Figure 2-44.

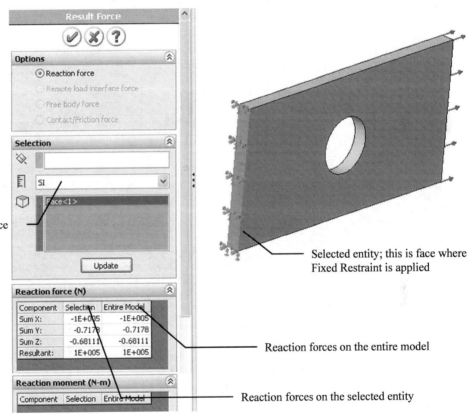

Figure 2-44: Result Force window

Reaction forces can also be displayed in components other than those defined by the global reference system. To do this, reference geometry such as plane or axis must be selected.

Notes:

3: Static analysis of an L-bracket

Topics covered

- Stress singularities
- Differences between modeling errors and discretization errors
- Using mesh controls
- Analysis in different SolidWorks configurations
- Nodal stresses, element stresses

Project description

An L-shaped bracket (in the file called L BRACKET in SolidWorks) is supported and loaded as shown in Figure 3-1. We wish to find the displacements and stresses caused by a 1,000 N bending load. In particular, we are interested in stresses in the corner where the 2-mm round edge (fillet) is located. Since the radius of the fillet is small compared to the overall size of the model, we decide to suppress it. As we will soon prove, suppressing the fillet is a bad mistake!

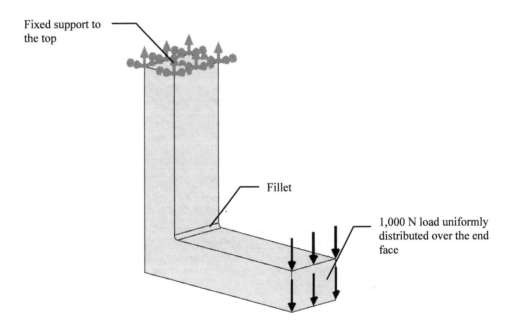

Figure 3-1: Loads and supports applied to the L BRACKET model

The geometry of the L BRACKET includes a fillet, which will be mistakenly suppressed, leaving in its place a sharp re-entrant corner.

The L BRACKET model has two configurations: *round corner* and *sharp corner*. Change to *sharp corner* configuration. Material (Alloy steel) is applied to the SolidWorks model and is automatically transferred to COSMOSWorks

Procedure

Following the same steps as those described in Chapter 2, define the study called *mesh 1* and define **Fixed** Restraint to the top face as shown in Figure 3-1. Define load using **Force** window choice shown in Figure 3-2.

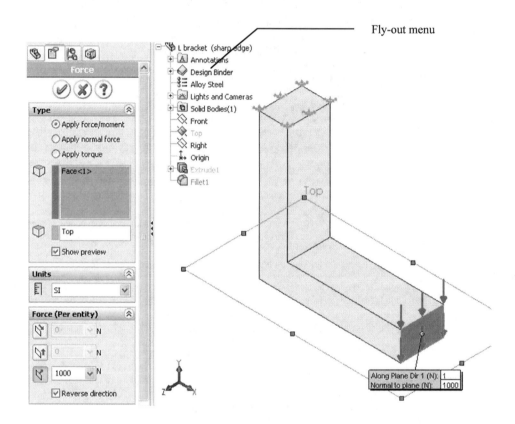

Figure 3-2: Force definition window uses Apply force/moment type of force. Force direction is specified as normal to the Top reference plane, which is used as a reference to determine force direction.

The reference plane can be conveniently selected from a "fly-out" SolidWorks menu.

Next, mesh the model with second order tetrahedral elements, accepting the default element size. The finite element mesh is shown in Figure 3-3.

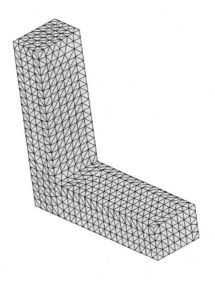

Figure 3-3: Finite element mesh created using the default setting of the automesher

In this mesh, the global element size is 4.76 mm.

The displacement and stress results obtained in the *mesh 1* study are shown in Figure 3-4.

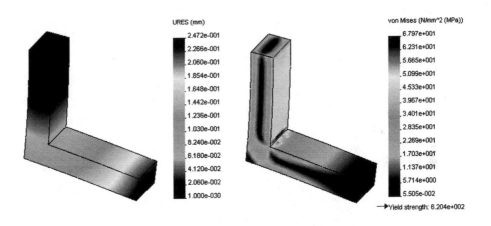

Figure 3-4: Displacements and von Mises stresses results produced using *mesh 1.*

The maximum displacement is 0.247 mm and the maximum von Mises stress is 68 MPa.

Now we will investigate how using smaller elements affects the results. In Chapter 2, we did this by refining the mesh uniformly so that the entire model was meshed with elements of a smaller size. Here we will use a different technique. Having noticed that the stress concentration is near the sharp re-entrant corner, we will refine the mesh locally in that area by applying mesh controls. The element size everywhere else will remain the same as it was before: 4.76mm.

Copy the *mesh1* study into the *mesh2* study. Notice that the name of the active study is always shown in bold. Select the edge where mesh controls will be applied, then right-click the *Mesh* folder in the *mesh2* study (this folder is empty at this moment) to display the pop-up menu shown in Figure 3-5.

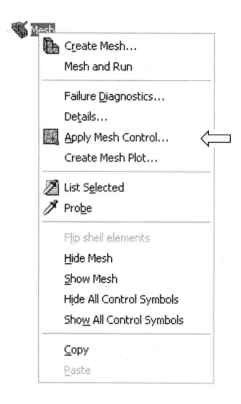

Figure 3-5: Mesh pop-up menu

Select **Apply Mesh Control…**, which opens the **Mesh Control** window (Figure 3-6). It is also possible to open the **Mesh Control** window first and then select the desired entity or entities (here the re-entrant edge) where mesh controls are being applied.

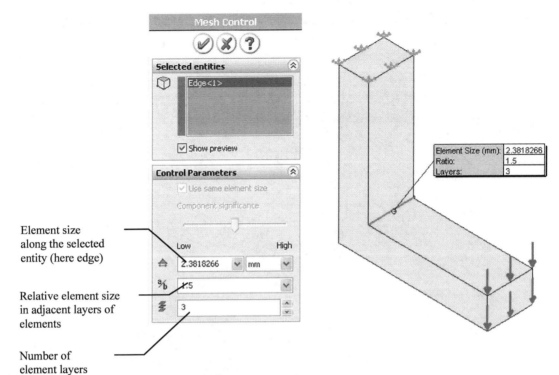

Element size along the selected entity (here edge)

Relative element size in adjacent layers of elements

Number of element layers affected by the applied control

Figure 3-6: Mesh Control window

Mesh controls allow you to define the local element size on selected entities. Accept the default values of the Mesh Control window.

The element size along the selected edge is now controlled independently of the global element size. Mesh control can also be applied to vertexes, faces and to entire components of assemblies. Having defined the mesh control, create a mesh with the same global element size as before (4.76 mm), while making elements along the specified edge to be 2.38 mm. The added mesh controls shows as *Control-1* icon in *Mesh* folder and can be edited using a pop-up menu displayed by right-clicking the *Mesh Control* icon (Figure 3-7).

Figure 3-7: Pop-up menu for the *Mesh Control* icon

The mesh with applied control (also called mesh bias) is shown in Figure 3-8.

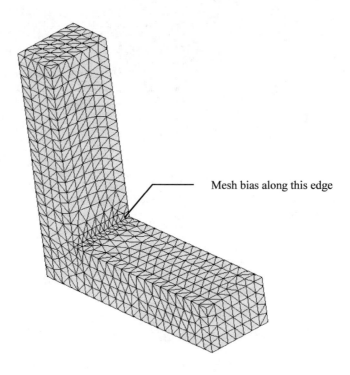

Figure 3-8: Mesh with applied controls (mesh bias)

Mesh 2 is refined along the selected edge. The effect of mesh bias extends for three layers of elements adjacent to the edge as specified in Mesh Control definition window (Figure 3-6).

The maximum displacements and stress results obtained in study *mesh 2* are 2.47697 mm and 74.1 MPa respectively. The number of digits shown in a result plot is controlled using **Chart Options** (right-click plot and select **Chart Options**).

Now repeat the same exercise three more times using progressively smaller elements along the sharp re-entrant edge (Figure 3-9):

- Study *mesh 3*: element size 1.19mm
- Study *mesh 4*: element size 0.60mm
- Study *mesh 5*: element size 0.30mm

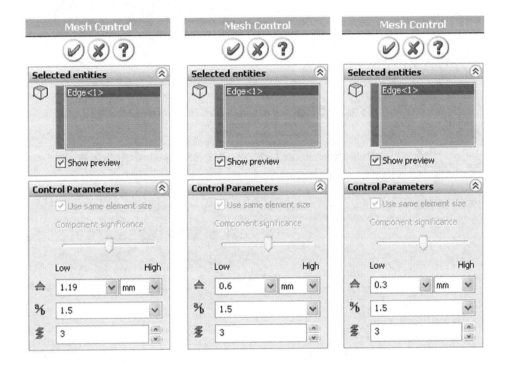

Figure 3-9: Mesh Control windows in studies *mesh 3*, *mesh 4*, and *mesh 5*

The summary of results of all five studies is shown in Figures 3-10 and 3-11.

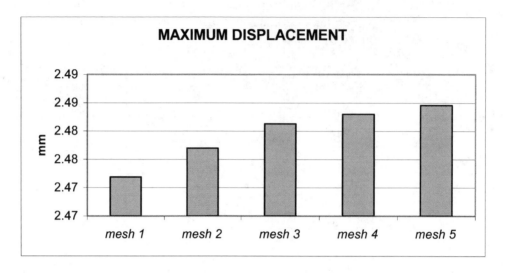

Figure 3-10: Summary of maximum displacement results

While each mesh refinement brings about an increase in the maximum displacement, the difference between consecutive results decreases.

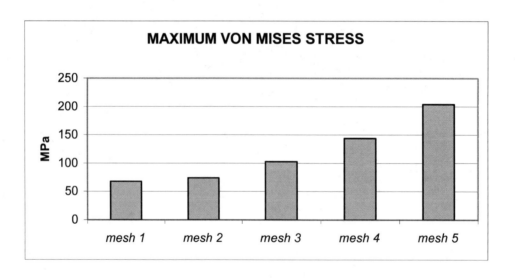

Figure 3-11: Summary of maximum stress results

Each mesh refinement brings about an increase in the maximum stress. The difference between consecutive results increases, proving that the maximum stress result is divergent.

We could continue with this exercise of progressive mesh refinement either:

❑ Locally, near the sharp re-entrant, as we have done here by means of mesh controls, or

❑ Globally, by reducing the global element size, as we did in Chapter 2.

We notice that displacement results converge to a finite value and that even the first mesh is good enough if we are looking only for displacements.

Stress results behave quite differently than displacement results. Each mesh refinement produces higher stress results. Instead of converging to a finite value, the maximum stress magnitude diverges. Given enough time and patience, we can produce results showing any stress magnitude we want. All that is necessary is to make the element size small enough!

The reason for divergent stress results is not that the finite element model is incorrect, but that the finite element model is based on the wrong mathematical model.

According to the theory of elasticity, stress in the sharp re-entrant corner is infinite. A mathematician would say that stress in the sharp re-entrant corner is singular. Stress results in a sharp re-entrant corner are completely dependent on mesh size: the smaller the element, the higher the stress. Therefore, we must repeat this exercise after un-suppressing the fillet, which is done in by changing configuration from *sharp edge* to *round edge* in SolidWorks Configuration Manager.

Notice that after we return from the Configuration Manager window to the COSMOSWorks window, all studies pertaining to the model in *sharp corner* configuration are not accessible. They can be accessed only if the model configuration is changed back to *sharp corner* (Figure 3-12).

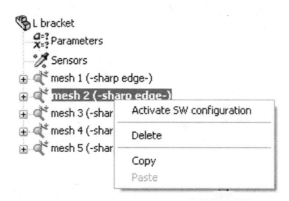

Figure 3-12: Studies become inaccessible when the model configuration is changed to a configuration other than that corresponding to now grayed-out studies

The SolidWorks model can be changed to a configuration corresponding to given study by right-clicking the study icon and selecting Activate SW configuration.

To analyze a bracket with fillet, define study *round corner*.

Since the fillet is a small feature compared to the overall size of the model, meshing with the default mesh settings will produce an abrupt change in element size between the fillet and the adjacent faces. To avoid this problem, we will try two different approaches. First select the **Automatic transition** option in the **Default Options** window (Figure 3-13).

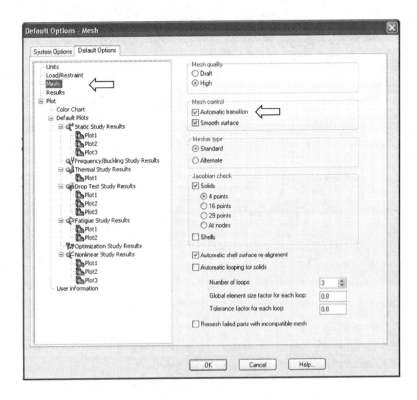

Figure 3-13: Meshing preferences with **Automatic transition** selected

However, meshing with the default element size and automatic transition option selected produces elements with excessive turn angle (Figure 3-14).

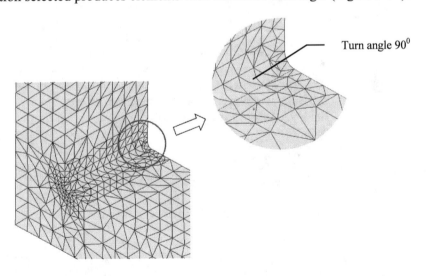

Figure 3-14: Mesh created with Automatic Transition selected features element with too high turn angle; here one element covers the angle of 90°.

It is recommended to use at least two elements to cover the 90°, meaning that an individual turn angle should not exceed 45°.

Therefore, in order to reduce the turn angle of elements meshing the fillet, we need to apply mesh control on the fillet face as well as use the **Automatic transition** (Figure 3-15).

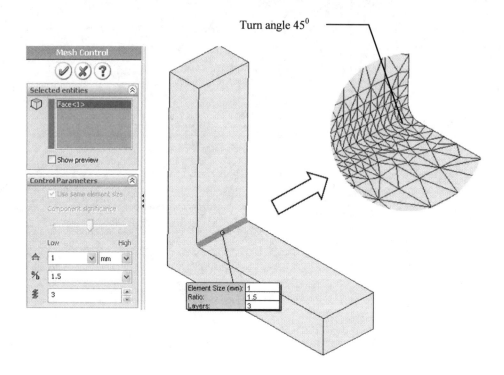

Figure 3-15: Mesh created with Automatic Transition and Mesh Control, shown on the left, has elements with turn angle of 45°

The L-BRACKET example is a good place to review the different ways of displaying stress results. Figure 3-16 shows the node values and element values of von Mises results produced in the study *round corner auto transition*. To select either node values or element values, right-click the *Plot* icon and select **Edit Definition**. This will open the **Stress Plot** window.

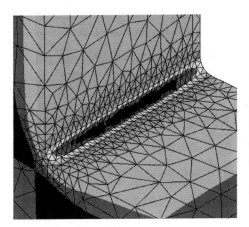 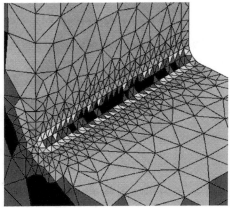

Figure 3-16: Von Mises stresses displayed as node values (left) and element values (right)

The irregularities in the shape of discrete fringes showing nodal stress results (left) may be used to decide if more mesh refinement is needed in the area of stress concentration. Here, we would recommend having elements with even lower turn angle. This could be accomplished by a more aggressive mesh control and/or by using small global element size.

As we explained in Chapter 1, nodal displacements are computed first. Strains and then stresses are calculated from the displacements results. Stresses are first calculated inside the element at certain locations, called Gauss points. Next, stress results are extrapolated to all of the elements' nodes. If one node belongs to more than one element (which is always the case unless it is a vertex node), then the stress results from all those elements sharing a given node are averaged and one stress value, called a node value, is reported for each node.

An alternate procedure to present stress results is by obtaining stresses in Gauss points, then averaging them in-between themselves. This means that one stress value is calculated for the element. This stress value is called an element value.

Node values are used more often because they offer smoothed out, continuous stress results. However, examination of element values provides important feedback on the quality of the results. If element values in two adjacent elements differ too much it indicates that the element size at this location is too large to properly model the stress gradient. By examining the element values, we can locate mesh deficiencies without running a convergence analysis.

To decide how much is "too much" of a difference requires some experience. As a general guideline, we can say that if the element values of stress in adjacent elements are apart by several colors on the default color chart, then a more refined mesh should be used.

Notes:

4: Stress and frequency analyses of a thin plate

Topics covered

- Use of shell elements
- Frequency analysis

Project description

We will analyze a support bracket (shown in Figure 4-1) with the objective of finding stresses and the first few modes of vibration. This will require running both static and frequency analyses. Open part model SUPPORT BRACKET with assigned material properties of AISI 304.

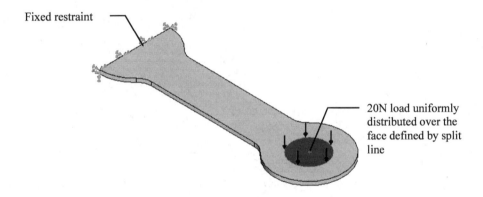

Figure 4-1: Support bracket

Note that CAD model contains split faces used by COSMOSWorks for load application.

Procedure

Before defining the study, consider that thin wall geometry would be difficult to mesh with solid elements. Generally it is recommended that two layers of second order tetrahedral elements be used across the thickness of a wall undergoing bending. Therefore, a large number of solid elements would be required to mesh this thin model.

Instead of using solid elements, we will use shell elements to mesh the surface located mid-plane in the bracket thickness. The study definition with the meshing option specifying **Shell mesh using mid-surfaces** is shown in Figure 4-2.

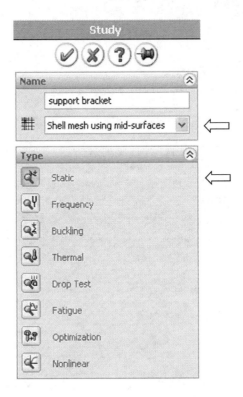

Figure 4-2: Study window and COSMOSWorks message

The study support bracket defines the Mesh type as a Shell mesh using mid-surfaces.

Before the study can be created, we must acknowledge the warning message that the option of creating shell elements using mid-surfaces works only for simple geometries (Figure 4-3).

Figure 4-3: A warning message

This is a reminder that the function of creating shell elements using mid-surface is still under development.

To apply loads, select the face shown in Figure 4-4 and apply a 20N normal force.

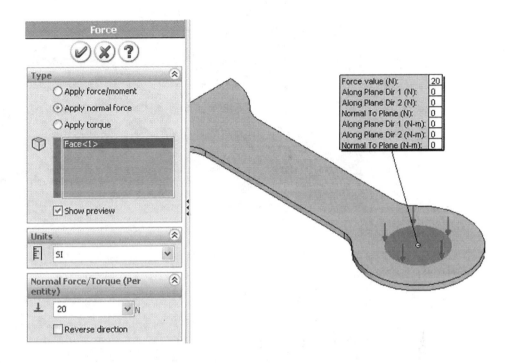

Figure 4-4: Force definition window

Support is applied to the end face, as shown in Figure 4-1. Since shell elements have six degrees of freedom, there is now a difference between applying fixed supports and immovable supports. In our case, we need a fixed support to eliminate all six degrees of freedom because an immovable support would only eliminate 3 degrees of freedom (translational), leaving rotational degrees of freedom unsupported. This would create an unintentional hinge in place of the intended fixed support.

The **Restraint** definition window is shown in Figure 4-5.

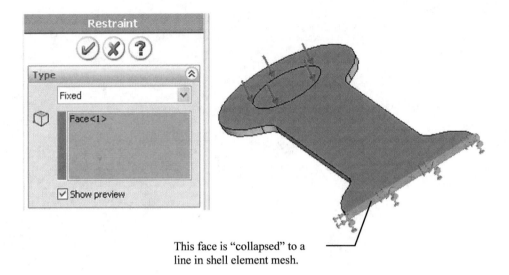

This face is "collapsed" to a line in shell element mesh.

Figure 4-5: Restraint window

Note that a Fixed support rather than Immovable support is selected. When using a Shell element mesh created in mid-plane, the shown face will reduce to a line and Immovable support would produce a hinge.

We do not explicitly define the shell thickness. COSMOSWorks assigns shell thickness automatically, based on the corresponding dimension of the solid CAD model, which in this case is 5mm.

The model is now ready for meshing (right-click *Mesh* folder, **Create Mesh**).

Shell element mesh is shown in Figure 4-6.

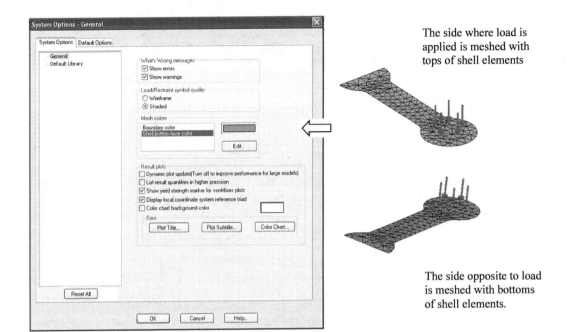

Figure 4-6: Shell element mesh. In the shell element mesh, elements have been placed mid surface between the faces that define the thin wall.

Different colors distinguish between the top and bottom of the shell elements. The bottom face color is specified in System Options window under Mesh options. The top face color is the same as the color of the SolidWorks model. Mesh shown in this illustration uses larger elements for better clarity of this illustration.

COSMOSWorks creates the mid-surface where shell elements are placed. In a SolidWorks model, this surface shows as an imported feature visible in the SolidWorks Feature Manager (Figure 4-7).

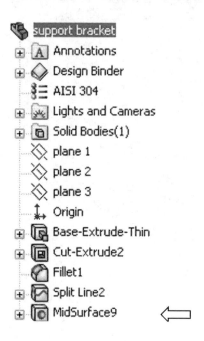

Figure 4-7: Mid-surface is an imported feature in the SolidWorks Manager

The surface meshed with shell elements has been automatically created by COSMOSWorks and appears as an imported feature in SolidWorks Feature Manager.

Review the mesh colors to ensure that the shell elements are aligned. Try reversing the shell element orientation: select the face where you want to reverse the orientation, right-click the *Mesh* folder to display a pop-up menu and select **Flip shell elements** (Figure 4-8).

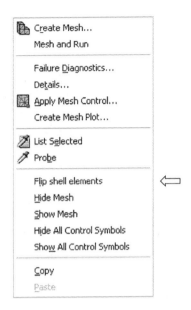

Figure 4-8: Pop-up menu for modifying shell element orientation

If desired, you may reverse shell element orientation with this menu choice. In this exercise reversing shell element orientation is not required.

Misaligned shell elements lead to the creation of erroneous plots like one shown in Figure 4-9, which shows a rectangular plate undergoing bending.

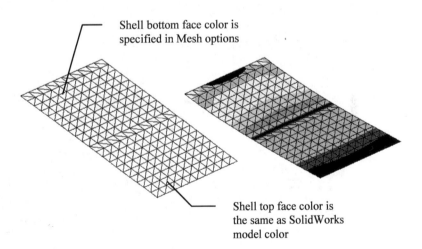

Figure 4-9: Misaligned shell elements and erroneous von Mises Stress plot resulting from this misalignment

The misaligned shell element mesh (left) and erroneous von Mises Stress plot are the result of shell element misalignment. This model is unrelated to our exercise.

Stress results on top and bottom sides of shell elements are shown in Figures 4-10 and 4-11.

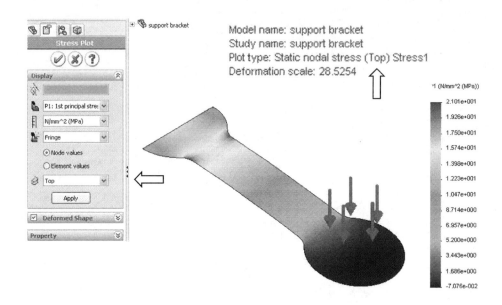

Figure 4-10: Maximum principle stress (P1) results for the top faces of the shell elements which is on the tensile side of the model

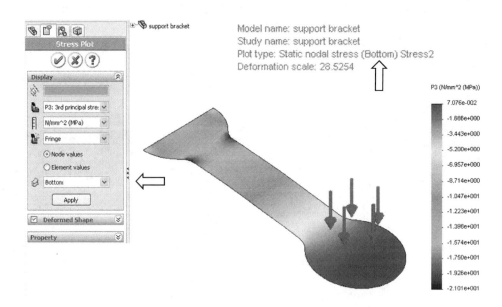

Figure 4-11: Minimum principle stress (P3) results for the bottom faces of the shell elements which is on the compressive side of the model

Note that although results pertain to the bottom side which is invisible, they can still be seen from the top, as in this illustration. We may say that the shell element model is "transparent" when reviewing results.

In Figure 4-11 we are looking at top faces of shell elements. Still, stress results are displayed for the bottom side (which is "underneath" the model) of the elements as if the shells were transparent. What stress is visible (top or bottom) does not depend on view direction (which side is visible) but only on the selection made in the **Stress Plot** window.

Having obtained displacement and stress solutions (we have not reviewed displacement results), the structural analysis of the support bracket has been completed. We now proceed to calculate natural frequencies for the same bracket. This requires a frequency analysis (Figure 4-12), also known as modal analysis.

Figure 4-12: Frequency study definition

You can copy restraint and mesh definitions from the static study to the frequency study by the dropping definition icon into the corresponding icon in the frequency study. No loads are defined in this frequency study. Figure 4-13 shows the result folders of frequency study.

Figure 4-13: The frequency study *(support bracket fr)* after solution

Deformation plots are automatically created for all modes of vibration as specified in the Options of Frequency study window (Figure 4-14).

Even though the study is complete, let us go back and review the options of a frequency study (right-click study icon, **Properties**). The **Frequency** window offers the choice of how many frequencies and associated modes of vibration will be calculated (Figure 4-14).

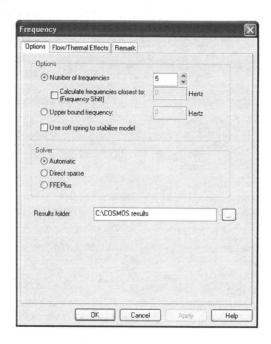

Figure 4-14: Frequency window

The Options tab in the Frequency window allows for the selection of the desired number of frequencies. By default, the first five frequencies are calculated.

By not modifying anything, we have accepted the default number of five frequencies. The **Deformation** plots shown in Figure 4-13 are five result plots corresponding to these five frequencies.

Figure 4-15 shows how a plot can be redefined to show a different mode.

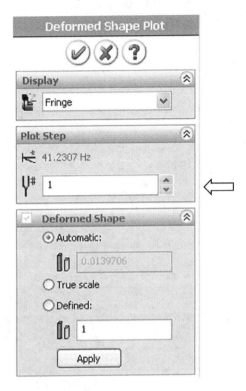

Figure 4-15: Displacement plot window

The definition of Deformed Shape Plot in the Frequency analysis requires us to specify the mode to be displayed. Here we select the first mode with a frequency of 41.2Hz. The magnitude of deformed shape is selected for the best visualization of the mode of vibration. It has no physical meaning.

Figure 4-16 shows displacement results related to the first mode of vibration. It is not created automatically but must be defined as shown in Figure 4-16.

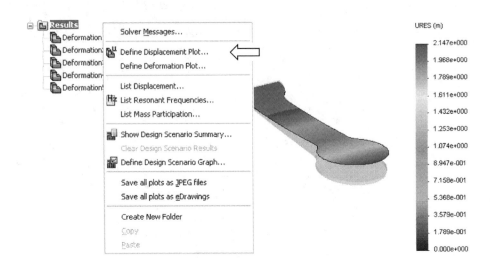

Figure 4-16: Displacements associated with *Mode 1*

Note that this undeformed shape is superimposed on the deformed shape. This option is selected in the Settings of result plot.

Even though the displacement plot (Figure 4-16) does show displacement magnitude, the displacement results have no quantitative importance in frequency analysis. Let us repeat this: Frequency analysis does not provide any quantitative information on displacements.

Displacement results are purely qualitative and can be used only for qualitative comparison of displacements within the same mode of vibration. Relative comparison of displacements between different modes is invalid.

Examine deformation plots of higher modes (Figure 4-17) to notice that higher modes are associated with more deformation. The best way to analyze the results of a frequency analysis is by examining the animated deformation plots.

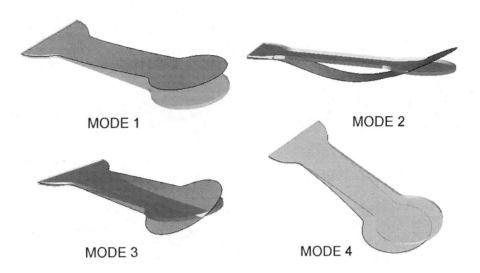

Figure 4-17: Deformation plots showing the shape of deformation (mode shape) of the first four modes.

COSMOSWorks results plots can be viewed more than one at a time using SolidWorks split window technique.

In the images shown in Figure 4-17 the undeformed shape appears as a solid and the deformed shape appears as a surface because shell elements are used in this study. To superimpose model on the deformed shape, make the proper selection in plot **Settings**.

To animate any plot, right-click a plot icon to display an associated pop-up menu, and then select **Animate**.

To **List Resonance Frequencies**, right-click the *Results* folder and make an appropriate selection (Figure 4-18).

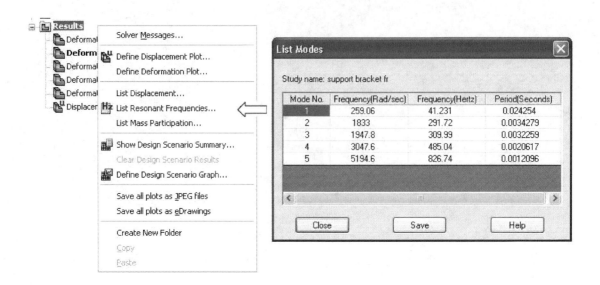

Figure 4-18: The summary of frequency results includes the list of all calculated modal (resonant) frequencies

5: Static analysis of a link

Topics covered

- Symmetry boundary conditions
- Defining restraints in a local coordinate system
- Preventing rigid body motions
- Limitations of small displacements theory

Project description

We need to calculate displacements and stresses of a link shown in Figure 5-1. The link is supported by tight-fitting pins in the two end holes and is loaded by a loose-fitting pin at the central hole with force of 100,000N. The other two holes are not loaded. Open part file LINK. It has assigned material properties of Chrome Stainless Steel.

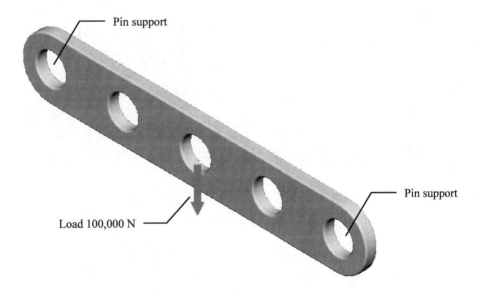

Figure 5-1: CAD model of the link

Note that the supporting pins and the loaded pin are not present in the model. All rounded edges have no structural significance and should be suppressed to simplify meshing.

Procedure

One way to conduct this analysis would be to model both the link and all three pins, and then conduct an analysis of assembly. However, we are not interested in the contact stresses that will develop between pins and the link. Our focus is on deflections and stresses that will develop in the link, so the analysis can be simplified. Instead of modeling the pins, we can simulate their effects by properly defining the restraints and the load. Notice that the link geometry, restraints, and loads are all symmetrical. We can take advantage of this symmetry and analyze only half of the model, replacing the other half with symmetry boundary conditions.

To work with half of the model, un-suppress the cut which is the last feature in the SolidWorks Feature Manager window. Also suppress all the small fillets in the model. The fillets have negligible structural effect and would unnecessarily complicate the mesh. Removing geometry details deemed unnecessary for analysis is called defeaturing.

Finally, note a split face in the middle hole that defines the area where load will be applied. Geometry in FEA-ready form is shown in Figure 5-2. Figure 5-2 also explains how restraints should be applied.

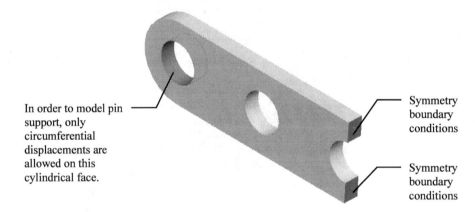

Figure 5-2: Half of the link with restraints explained

The applied restraints: a hole where the pin support is simulated and two faces in the plane of symmetry where symmetry boundary conditions are required.

The model is ready for the definition of supports – the highlight of this exercise. Move to COSMOSWorks and define a study as static analysis with solid elements.

Select the pin supported cylindrical face, as shown in Figure 5-2 and open the **Restraint** window (Figure 5-3).

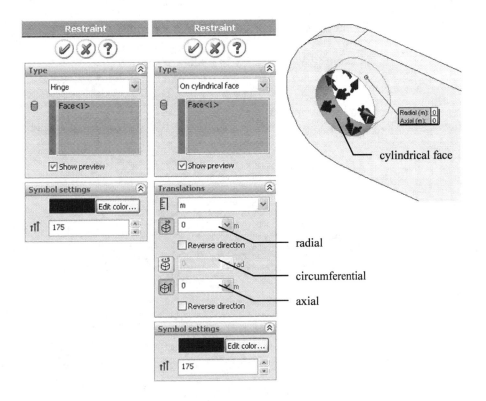

Figure 5-3: Restraint window

Identical results can be obtained using either a Hinge or On cylindrical face (with suppressed radial and axial directions) type of restraint.

When the **Restraint** definition window specifies the restraint type as **Hinge** or **On cylindrical face**, the restraint directions are associated with the directions of cylindrical face (radial, circumferential, and axial), rather than with global directions x, y, z.

To simulate the pin support that allows the link to rotate about the pin axis, radial displacement needs to be restrained and circumferential displacement allowed. Furthermore, displacement the in axial direction needs to be restrained in order to avoid rigid body motions of the entire link along this direction.

Functionally, the same restraint can be defined as **Hinge** support. Using **Hinge** support is faster and easier but has less learning benefit. This is why we use the **On cylindrical face** restraint.

Notice that while we must restrict the rigid body motion of the link in the direction defined by the pin axis, we can do this by restraining any point of the model. It is simply convenient to remove rigid body motions by applying the axial restraints to this cylindrical face.

To simulate the entire link, even though only half of geometry is present, we apply symmetry boundary conditions to the two faces located in the plane of symmetry. Symmetry boundary conditions allow only in-plane displacements. The easiest way to define symmetry boundary conditions is to use **Symmetry** as a type of restraint. The definition of the symmetry boundary conditions is illustrated in Figure 5-4.

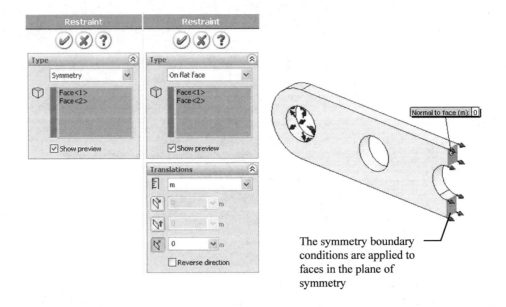

Figure 5-4: Definition of symmetry boundary conditions

The same could be defined using either a Symmetry restraint or, for example, On flat face restraint, where both in-plane displacements are allowed, but displacement in the direction normal to face is not allowed (set to zero).

Recall from Figure 5-1 that the link is loaded with 100,000 N. Since we are modeling half of the link, we must apply a 50,000 N load to a portion of the cylindrical face, as shown in Figure 5-5. The size of the load application area is arbitrarily created with a split line. It should be close to what we expect the contact area to be between the loose-fitting pin and the link.

When defining the load (Figure 5-5), take advantage of SolidWorks' fly-out menu visible in COSMOSWorks window to select reference plane required in **Force** definition window.

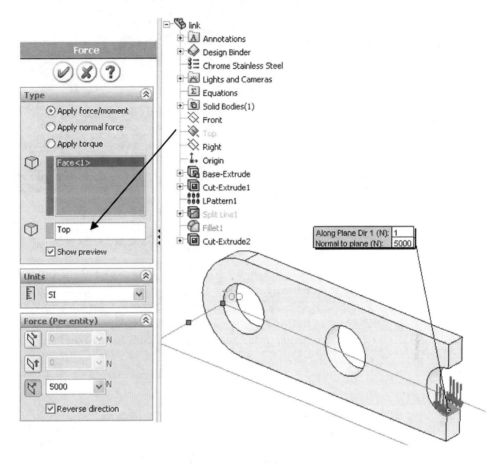

Figure 5-5: 50,000N force applied to the central hole

The Top reference plane is used to determine the load direction. Notice that the load is distributed uniformly. We are not trying to simulate a contact stress problem.

The last task of model preparation is meshing. Right-click the *Mesh* folder to display the related pop-up menu, and then select **Create....** Verify that the mesh preferences are set on high quality (meaning that second order elements will be created) and mesh the geometry using the default element size. For more information on the created mesh, you may wish to review **Mesh Details** (Figure 5-6).

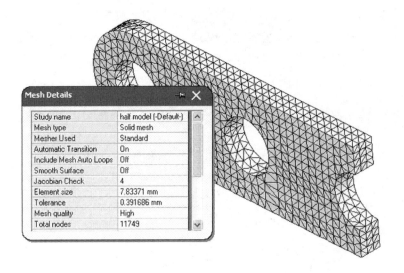

Figure 5-6: Mesh Details window shown over the meshed model

After solving the model, we first need to check if the pin support and symmetry boundary conditions have been applied properly. This includes checking whether the link can rotate around the pin and whether it behaves as a half of the whole link. This is best done by examining the animated displacements, preferably with both undeformed and deformed shapes visible (Figure 5-7).

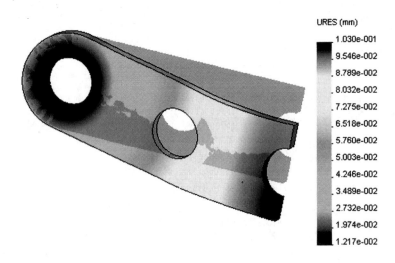

Figure 5-7: Comparison of the deformed and undeformed shapes

Comparison of the deformed and undeformed shapes verifies the correctness of the restraints definition: the link rotates around the imaginary pin while faces in the plane of symmetry remain flat and perform only in-plane translations.

To conclude this exercise, review the stress results. Examine the different stress components, including the maximum principal stresses, minimum principal stresses, etc.

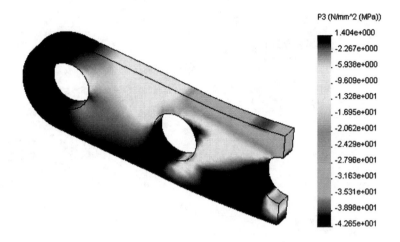

Figure 5-8: Sample of stress results

In this plot, the minimum principal stress (P3) plot is shown.

Repeat this exercise using the full model. To do this, open the SolidWorks Manager, suppress the cut, and return to COSMOSWorks to perform analysis of complete model without using symmetry boundary conditions.

Before finishing the analysis of LINK, we should notice that the link supported by two pins as modeled in this exercise corresponds to the configuration shown in Figure 5-9, where one of the hinges is free to move horizontally.

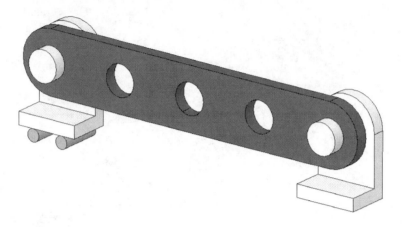

Figure 5-9: Our model corresponds to the situation where one hinge is floating, symbolically shown here by rollers under the left hinge

Since linear analysis does not account for changes in model stiffness during the deformation process, linear analysis is unable to model stresses that would have developed if both pins were in a fixed position. If both pins were fixed, a non-linear geometry analysis would be required to analyze the model. Refer to Chapter 17 for more information.

6: Frequency analysis of a tuning fork

Topics covered

- Frequency analysis with and without supports
- Rigid body modes
- The role of supports in frequency analysis
- Symmetric and anti-symmetric modes

Project description

Structures have preferred frequencies of vibration, called resonant frequencies. A mode of vibration is the shape in which a structure will vibrate at a given natural frequency. The only factor controlling the amplitude of vibration in resonance is damping. While any structure has an infinite number of resonant frequencies and associated modes of vibration, only a few of the lowest modes are important to describe their response to dynamic loading. A frequency analysis calculates these resonant frequencies and their associated modes of vibration.

Open the part file called TUNING FORK. It has material properties already assigned (Chrome Stainless Steel). The model is shown in Figure 6-1.

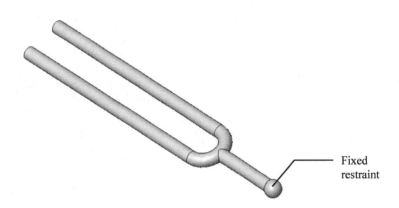

Figure 6-1: Tuning fork geometry

Fixed restraint is applied to the surface of the ball. Do not apply an On spherical face restraint.

A quick inspection of the CAD geometry reveals a sharp re-entrant edge. This condition renders the geometry unsuitable for stress analysis, but still acceptable for frequency analysis.

Procedure

Define **Frequency** study as shown in Figure 6-2.

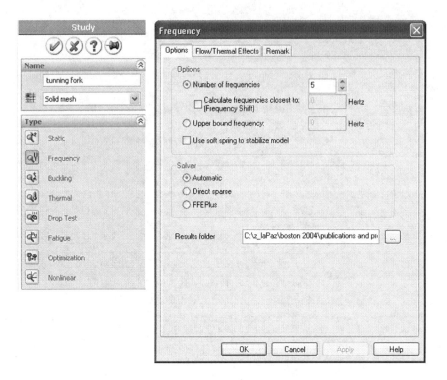

Figure 6-2: Frequency study definition (left) and study properties (right)

The Options tab in the Frequency window allows for specifying the number of frequencies to be calculated in the frequency study. Notice that, as always, different solvers are available. We request that five frequencies be calculated using the automatically selected solver.

Next, define fixed restraints to the ball surface, as shown in Figure 6-1. This approximates the situation when the tuning fork is held with two fingers.

Finally, mesh the model with the default element size. The meshed model is shown in Figure 6-3. The automesher selects the element size to satisfy the requirements of a stress analysis. A frequency analysis is less demanding on the mesh, so generally, a less refined mesh is acceptable. Nevertheless, since this is a very simple model, we accept the mesh without making an attempt to simplify it.

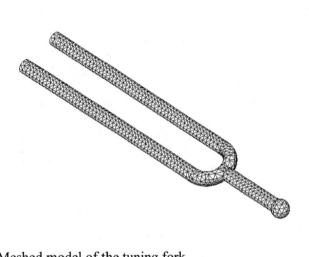

Figure 6-3: Meshed model of the tuning fork

After the solution is complete, COSMOSWorks automatically creates deformation plots (Figure 6-4). For reasons already explained in Chapter 4 we won't even look at displacement results.

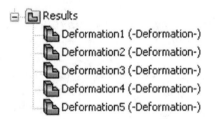

Figure 6-4: Five Deformation plots are created automatically

Plots Deformation1 shows Mode Shape 1, etc.

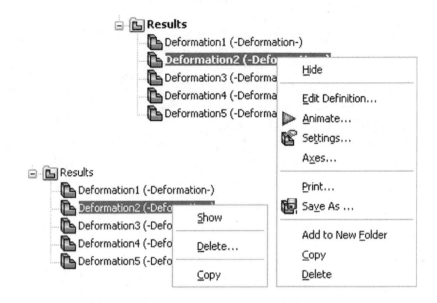

Figure 6-5: Displaying and modifying plots

The plot is displayed by double-clicking or right-clicking to activate pop up menu and selecting, Show (left). Plot can be modified using selections from the pop-up menu activated by right-clicking the plot icon (right).

The deformation plot in the modal analysis shows the associated modal shape of vibration and lists the corresponding natural frequency. The first four modes of vibration are presented in Figure 6-6.

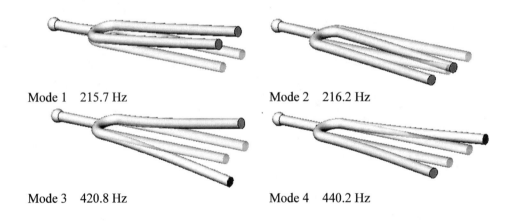

Mode 1   215.7 Hz   Mode 2   216.2 Hz

Mode 3   420.8 Hz   Mode 4   440.2 Hz

Figure 6-6: First four modes of vibration and their associated frequencies

Plots also show the model superimposed on the deformed shape as selected in the Settings of the plot.

The analyzed tuning fork is the most common type of tuning fork and, as any musician will tell us, should produce a lower A sound, with frequency of 440 Hz.

However, the lower A frequency of 440 Hz, which we were expecting to be the first mode, is actually the fourth mode. Before explaining the reasons for this, let's run the frequency analysis again, this time without any restraints. We need to define a new frequency study, which we will call *tuning fork no supports*.

The easiest way is to do this is to copy the existing *tuning fork* study and either delete or suppress the restraint (right-click the restraint icon and make proper selection).

In the **Options** tab of the Frequency window in the *tuning fork no supports* study, we again specify that five modes be calculated (no change compared to previous study). After the solution is has been completed, right-click the *Deformation* folder and select **List Resonant Frequencies**.

Here, we find that the highest calculated mode is mode 11 (Figure 6-7) even though we asked that only five modes be calculated.

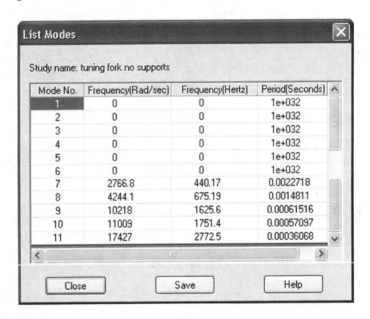

Figure 6-7: **List Modes** window

The illustration has been modified in a graphic program to show all eleven modes without scrolling.

We also notice that the first six modes have the associated frequency 0 Hz. Why? The first six modes of vibration correspond to rigid body modes. Because the tuning fork is not supported, it has six degrees of freedom as a rigid body: three translations and three rotations.

COSMOSWorks detects those rigid body modes and assigns them zero frequency (0 Hz). The first elastic mode of vibration, meaning the first mode requiring the fork to deform is mode 7, which has a frequency of 440.2 Hz. This is close to what we were expecting to find as the fundamental mode of vibration for the tuning fork.

Why did the frequency analysis with the restraint not produce the first mode with a frequency near to 440 Hz? If we closely examine the first three modes of vibration of the supported tuning fork, we notice that they all need the support in order to exist. The support is needed to sustain these modes, but while this support makes these first three modes possible, in reality it also provides damping. After modes 1, 2, and 3 have been damped out, the tuning fork vibrates the way it was designed to: in mode 4 (calculated in the analysis with supports) or mode 7 (calculated in the analysis without supports). These two modes are identical.

To learn more about modes of vibration of an unsupported elastic model (all models in FEA are considered elastic) calculate six elastic modes of an unsupported model EC027. Deformation results corresponding to the first six modes are shown in Figure 6-8.

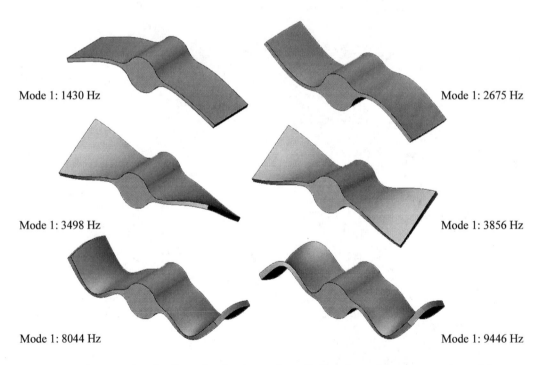

Mode 1: 1430 Hz

Mode 1: 2675 Hz

Mode 1: 3498 Hz

Mode 1: 3856 Hz

Mode 1: 8044 Hz

Mode 1: 9446 Hz

Figure 6-8: The first six elastic modes of vibration of unsupported model EC027

Based on deformation results of SUPPORT BRACKET, TUNING FORK and EC027 we can make an interesting observation about the nature of modes of vibration. If a model is symmetric and has symmetric restraints (or no restraint at all), then its modal shapes will be either symmetric or anti-symmetric.

7: Thermal analysis of a pipeline component and a heater

Topics covered

- Steady state thermal analysis
- Analogies between structural and thermal analysis
- Analysis of temperature distribution and heat flux

Project description

So far, we have performed static analyses and frequency analyses, which both belong to the class of structural analyses. Static analysis provides results in the form of displacements, strains, and stresses, while frequency analysis provides results in the form of natural frequencies and associated modes of vibration. We will now examine a thermal analysis. Numerous analogies exist between thermal and structural analyses. The most direct analogies are summarized in Figure 7-1.

Structural Analysis	**Thermal Analysis**
Displacement [m]	Temperature [K]
Strain [1] (dimensionless)	Temperature gradient [K/m]
Stress [N/m^2]	Heat flux [W/m^2]
Load [N] [N/m] [N/m^2] [N/m^3]	Heat source [W] [W/m] [W/m^2] [W/m^3]
Prescribed displacement [m]	Prescribed temperature [K]

Figure 7-1: Selected analogies between structural and thermal analysis

In this exercise, we will perform a thermal analysis of two parts.

Procedure

Open part model CROSSING PIPES. Our objective is to find the steady state temperature of the part when prescribed temperatures are applied to end faces as shown is Figure 7-2. As indicated in Figure 7-1, prescribed temperatures are analogous to prescribed displacements in structural analyses.

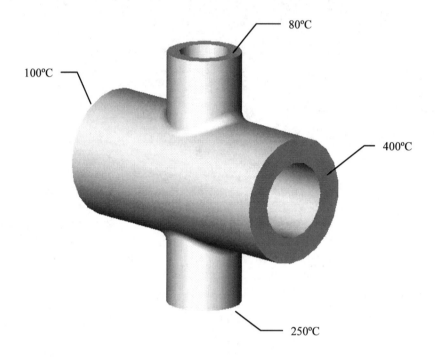

Figure 7-2: CAD model of two crossing pipes

Also shown are the prescribed temperatures, applied to the end faces as temperature boundary conditions.

Since no convection coefficients are defined on any faces, heat can enter and leave the model only through the end faces with prescribed temperatures assigned. Even though the problem has little relevance to real life heat transfer problems, it is a good example to help us understand the basics of thermal analysis.

The first step is study definition. Call this study *crossing pipes* and define it as shown in Figure 7-3.

Figure 7-3: Definition of *crossing pipes* thermal study

The study definition of crossing pipes is a Thermal analysis type using a Solid mesh.

To define the prescribed temperature, open the pop-up menu by right-clicking the *Loads/Restraints* folder and selecting **Temperature** (Figure 7-4).

Figure 7-4: Pop-up menu related to a thermal analysis

The pop-up menu lists the thermal boundary conditions (Temperature, Convection, Radiation) and thermal loads (Heat Flux, Heat Power) available in a thermal analysis.

Figure 7-5 shows the **Temperature** window for the face where temperature of 100°C is applied. Since each of the four faces has a different prescribed temperature, we need to assign temperatures in four separate steps.

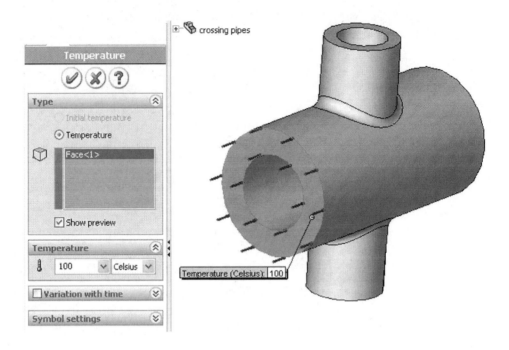

Figure 7-5: Temperature window

Note that the units are in degrees Celsius. Fahrenheit or Kelvin degrees can be used as well.

The next step is meshing the model. To ensure that fillets are meshed with elements with low turn angles, define the mesh control as shown in Figure 7-6. It is important for heat flux calculations that an element has a low turn angle of 30 degrees or less.

Use the default global element size and no **Automatic transition** to create the mesh shown in Figure 7-7.

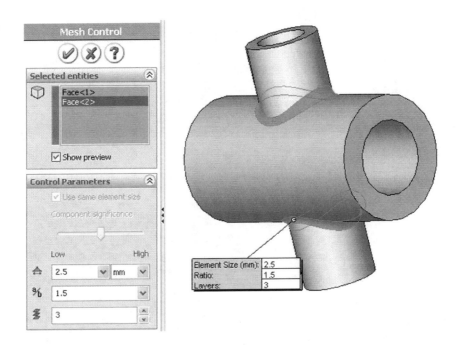

Figure 7-6: Mesh Control window specifies an element size 2.5mm on both fillets.

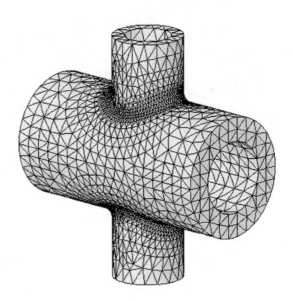

Figure 7-7: Finite element mesh of the crossing pipes model created with the default global element size and mesh control applied to both fillets.

After solving the model, notice that only one result folder called *Thermal* is present and has one plot. By default, it shows the temperature distribution. We need to create one more plot showing resultant heat flux (Figure 7-8).

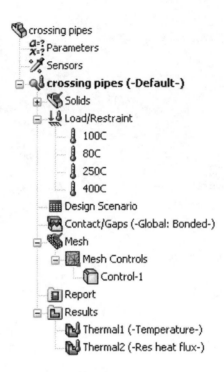

Figure 7-8: Crossing pipes study folders

The prescribed temperature definition icons have been renamed using the click-inside technique to give them more descriptive names. Resultant heat flux plot has been added to the Results folder.

Temperature and resultant heat flux plots are shown in Figure 7-9 and Figure 7-10.

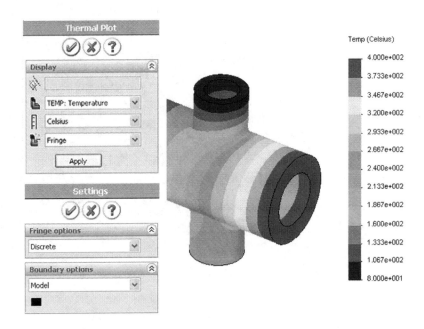

Figure 7-9: **Thermal Plot** window defining temperature plot and the corresponding result plot

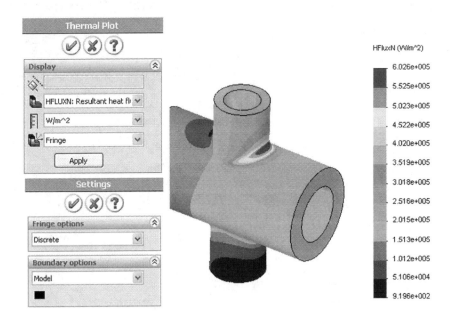

Figure 7-10: **Thermal Plot** window defining heat flux plot and the corresponding result plot

We will now conduct thermal analysis of a pipe with cooling fins. The objective of this analysis is to find how much heat dissipated by 50mm long section (Figure 7-11). Open part model HEATER.

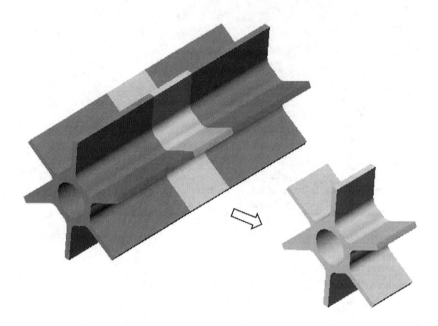

Figure 7-11: Analysis of pipe with cooling fins (left) is conducted on the model of a section (right)

Hot water at 100°C flows inside the heater. The coefficient of thermal convection between water and the heater face is 1000 W/(m²/K), meaning that each 1m² of the surface exchanges (gains or loses) 1W heat if the temperature difference between the face and water is 1K. The coefficient of thermal convection between outside faces and air is 20 W/(m²/K). The ambient air temperature is 27 °C or 300K. Define convection coefficients and bulk temperatures as shown in Figures 7-12 and 7-13.

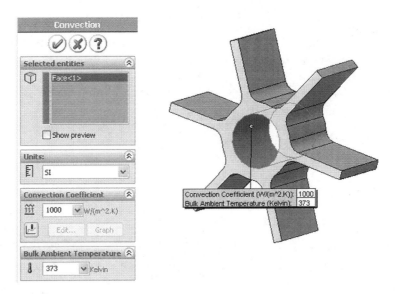

Figure 7-12: Convection coefficient and bulk temperature on the water side.

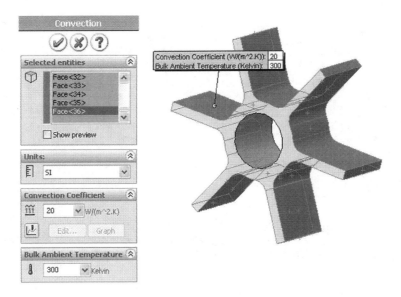

Figure 7-13: Convection coefficient and bulk temperature on the air side.

Because the model represents a section of a longer pipe, there is no heat exchange through the end faces. Therefore, we do not define any convection coefficients on the end faces, treating them as insulated.

Mesh the model with a default mesh size; solve and construct resultant heat flux plot. Right-click the plot icon and select **List selected** for the pop up menu to open the **Probe Results** window. In the model window, select the inside face, then click the **Update** button in the Probe results window. The total heat entering the model through the selected face is shown in the lower portion of **Probe Results** window (Figure 7-14).

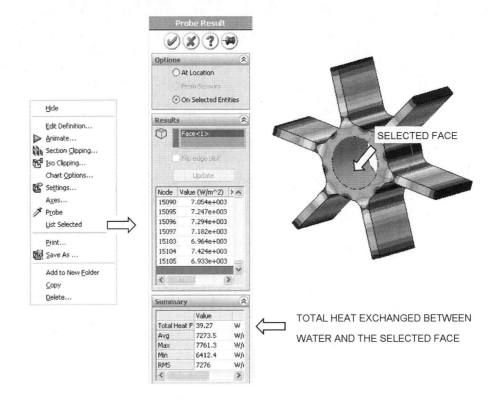

Figure 7-14: Total exchanged between water and model is 39.3W.

The positive sign indicates that the model gains heat from water. This can be visualized by constructing a vector plot of resultant heat flux.

Alternatively we can obtain the same results by selecting all faces on the air side (Figure 7-15). Selection of multiple entities requires pressing and holding the Ctrl key and is somewhat difficult because of poor visual feedback on the plot.

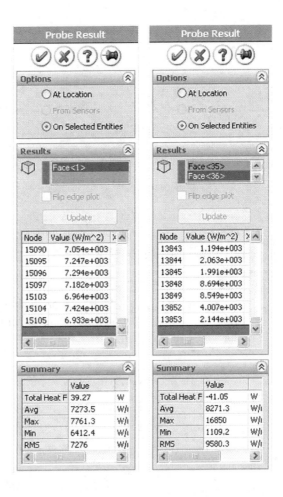

Figure 7-15: Total exchanged between water and model (left) and between model and air (right).

Heat gained from water is +39.3W. Heat lost by model to air is -41W.

Since this is steady state thermal analysis, the amount of heat entering the model and the amount of hear dissipated by the model must be equal. Any small discrepancies are due to numerical error.

The analysis can be significantly simplified by noticing that heat flow through the model has the property of axial symmetry. Therefore, instead of analyzing the entire model, we can analyze only one radial section (Figure 7-16).

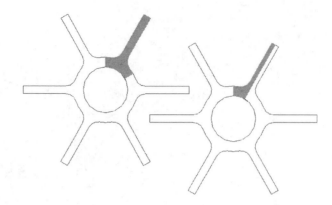

Figure 7-16: The model can be simplified to a radial section. While a 60° section (left) appears the most obvious at he first sight, the smallest repeatable section is 30° (right)

You are encouraged to repeat the analysis in model configuration *30 degrees section* or *60 degrees section*. Do not apply any convection conditions on radial faces and remember to multiply the total heat either by 6 or by 12, depending on which configuration you use for the analysis (Figure 7-17).

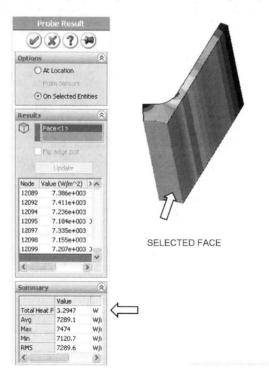

Figure 7-15: Total exchanged between water and the selected face of 30° section.

Total heat gained from water by model is 3.3 x 12 = 39.6W.

8: Thermal analysis of a heat sink

Topics covered

- Analysis of an assembly
- Global and local Contact/Gaps conditions
- Steady state thermal analysis
- Transient thermal analysis
- Thermal resistance layer
- Use of section views in result plots

Project description

In this exercise, we continue with thermal analysis. However, this time we will analyze an assembly rather than a single part. Open the assembly HEAT SINK (Figure 8-1).

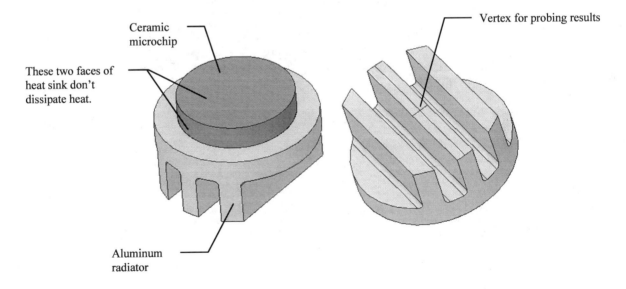

Figure 8-1: CAD model of a radiator assembly

The CAD model of a radiator assembly consists of two components: an aluminum radiator and a ceramic microchip.

Analysis of an assembly allows assignment of different material properties to each assembly component. Notice that the *Solids* folder contains two icons corresponding to the two assembly components with material properties already assigned. This is because material has been assigned to parts which are assembly components: **Ceramic Porcelain** material to the microchip and **1060 Alloy** to the radiator.

The ceramic insert generates a heat power of 25W and the aluminum radiator dissipates this heat. The ambient temperature is 27°C (300K). Heat is dissipated to the environment by convection through all exposed faces of the radiator. We assume that the microchip is insulated, meaning it can not dissipate heat directly to ambient air, but only through face touching the radiator. The convection coefficient (also called the film coefficient) is assumed to be $25W/m^2/K$ in this model. This means that if the difference of temperature between the face of the radiator and the surrounding air is 1K, then each square meter of the radiator surface dissipates 25W of heat. This value of the convection coefficient corresponds to natural convection without a cooling fan.

Heat flowing from the microchip to the radiator encounters thermal resistance on the boundary between microchip and radiator. Therefore, a thermal resistance layer must be defined on the interface between these two components.

Our first objective is to determine the temperature and heat flux of the assembly in steady state conditions, after enough time has passed for temperatures to stabilize. This will require steady state thermal analysis.

The second objective is to study the temperature in the assembly as a function of time in a transient process when the assembly is initially at room temperature and power is turned on at time t=0. This will require transient thermal analysis.

Procedure

Create a thermal study called *heat sink steady state*. Before proceeding, we need to investigate icon called *Contact/Gaps*, which is found in the COSMOSWorks Manager window. Right-click the *Contact/Gaps* icon to open the pop-up menu, shown in Figure 8-2.

Figure 8-2: *Contact/Gaps* icon in the COSMOSWorks Manager

By default, Global Contact is set as Bonded. As the name implies, all parts in assembly behave as one. We need to change it by defining a Contact Set.

We need to define a local **Contact Set** for contacting faces in order to introduce a thermal resistance between the contacting faces. This can only be done as local *Contact/Gaps* condition.

Right-click *Contact/Gaps* folder and select **Define Contact Set** (Figure 8-2) to open the **Contact Set** window shown in Figure 8-3. Select **Node to surface** as contact type. Referring to Figure 8-3, select contacting faces and enter **Distributed Thermal Resistance** as $0.0001 Km^2/W$ (this value is usually obtained by testing). Note that the units of thermal resistance are the reciprocal of the units of thermal convection.

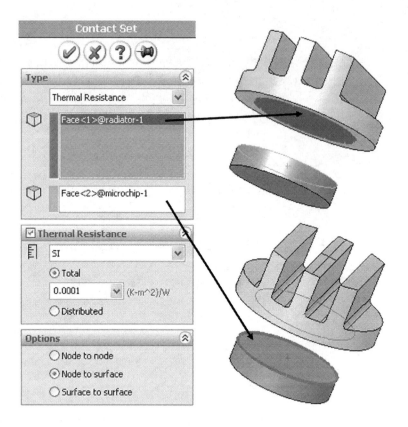

Figure 8-3: Defining thermal resistance between the microchip and radiator.

Modeling thermal resistance between contacting faces requires that contact between these two faces is defined as Thermal Resistance. Mesh options are Node to surface. This contact condition overrides the global Bonded contact condition.

Next, we specify the heat power generated in the microchip. To do this, right-click the *Load/Restraint* folder to open the pop-up menu. Now select **Heat Power...** to open the **Heat Power** window (Figure 8-4), then from fly-out menu select *microchip* assembly component. This applies heat power to the entire volume of the selected component.

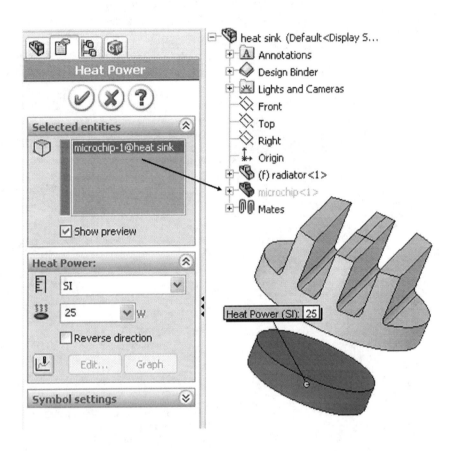

Figure 8-4: The SolidWorks pop-up menu is used to make a selection of the part (here a microchip) necessary to define Heat Power window

Right-click the Load/Restraint folder to open a pop-up menu. From this menu, select Heat Power to open Heat Power definition window. Notice that microchip-1 appears in the Selected Entity field.

So far we have assigned material properties to each component as well as defined the heat source and thermal resistance layer. In order for heat to flow, we must also establish a mechanism for the heat to escape the model. This is accomplished by defining convection coefficients.

Right-click the *Load/Restraint* folder to open a pop-up menu and select **Convection...** to open the **Convection** window (Figure 8-5). Select all faces of the radiator except the one touching the microchip. Enter 25 W/(m²K) as the value of convection coefficient for all selected faces and enter the ambient temperature (bulk temperature) as 300K.

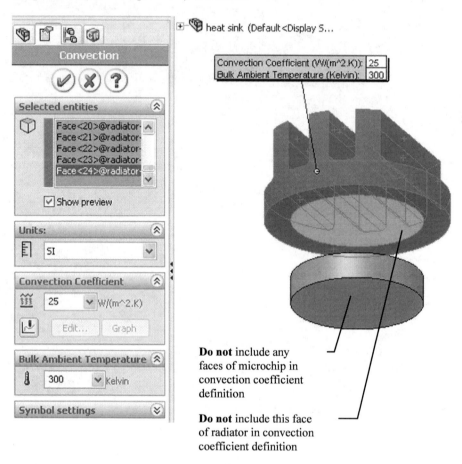

Do not include any faces of microchip in convection coefficient definition

Do not include this face of radiator in convection coefficient definition

Figure 8-5: Convection window

Use this window to specify both the Convection Coefficient and the Bulk Temperature.

The last step before solving is creating the mesh. For accurate heat flux results apply mesh control to all six fillets as shown in Figure 8-6.

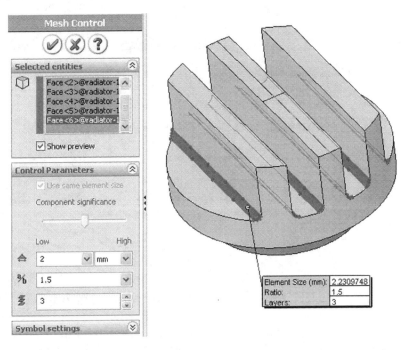

Figure 8-6: Mesh control applied to all fillets in the Radiator component

Now mesh the assembly with the default global element size to produce a mesh, as shown in Figure 8-7.

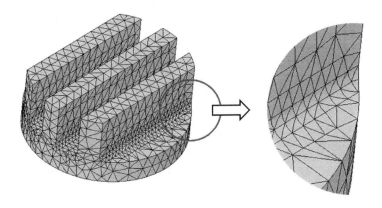

Figure 8-7: View of the meshed assembly

Note that the use of Mesh Controls allows us to have two elements meshing the 90° fillet. Consequently, the element turn angle is 45°. A turn angle of 90° is not recommended. Automatic transition is not used in this mesh.

Once the solution is ready, examine two plots: temperature and resultant heat flux. The temperature plot is created automatically in the *Results* folder. The required choices for both plots are shown in Figure 8-8.

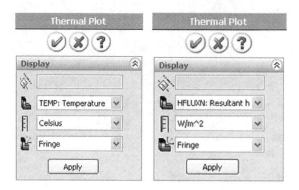

Figure 8-8: Thermal Plot definition window for temperature distribution plot (left) and heat flux plot (right)

Both temperature and heat flux result plots are more informative if presented using section views. COSMOSWorks offers a multitude of options for sections plots which are easier to practice than to explain. Here we describe the procedure of creating a flat section result plot using one of SolidWorks default reference planes (any reference plane can be used).

To show the section view of temperature distribution plot, right-click the plot icon and select **Section Clipping** from the pop-up menu. By default, the cutting surface is flat and aligned with the first reference plane in SolidWorks Feature Manager. To select another cutting plane, select it from the SolidWorks fly-out menu.

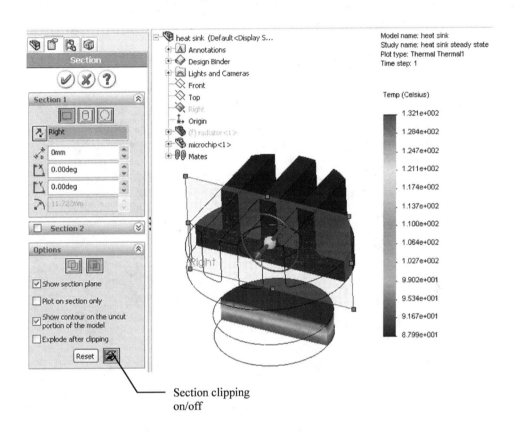

Figure 8-9: Section plot of temperature distribution in the assembly

Cutting plane position can be modified in a way similar to modifying a SolidWorks exploded view. SolidWorks exploded view is also used.

Experiment with different options in **Section** window, and then construct a section plot of heat flux. In addition to section clipping, use exploded view to produce a heat flux plot similar to one shown in Figure 8-10.

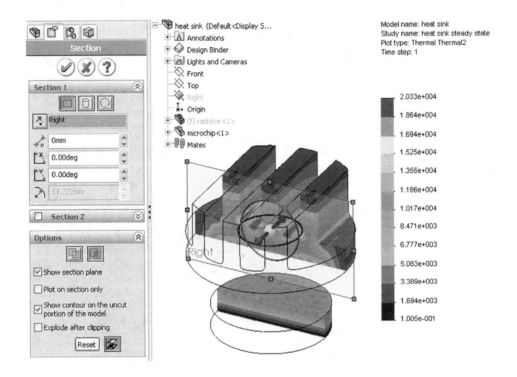

Figure 8-10: Section plot of heat flux in the assembly using exploded view.

Since heat flux is a vector quantity, heat flux results lend themselves well to representation by a vector type plot. Menu selections necessary to create and modify a vector type plot are shown in Figure 8-11.

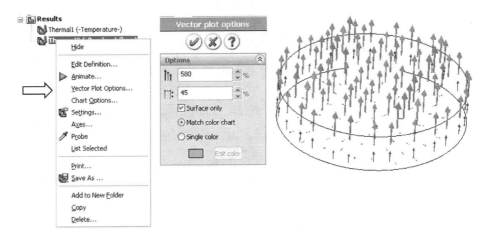

Figure 8-11: Vector plot of heat flux in the microchip

Vector plot shows only the microchip. The radiator component of the assembly is hidden using a SolidWorks function. Note that arrows "coming out" of the microchip visualize where heat leaves the microchip. Vectors are tangent to faces where no convection coefficients have been defined.

This completes the steady state thermal analysis of the heat sink assembly. We now proceed with transient thermal analysis. Copy study *heat sink steady state* into study *heat sink transient*. Right-click the study *heat sink transient* folder and select **Properties** to open the window shown in Figure 8-12.

Select **Transient** analysis (**Steady state** is the default option). Our objective is to monitor the temperature changes every 60 seconds during the first 3600 seconds, with particular attention to the vertex location shown in Figure 8-1. Enter 3600 as **Total time** and 60 as **Time increment**.

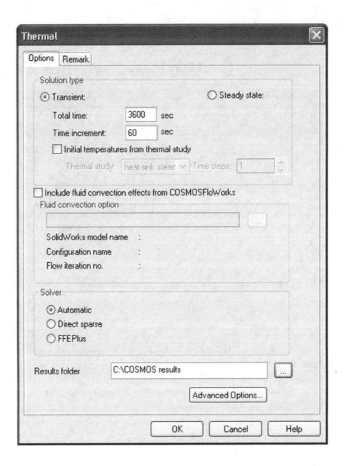

Figure 8-12: **Transient** thermal analysis is specified in thermal study **Options**

Analysis will be carried on for 3600 seconds with results reported every 60 seconds.

Transient thermal analysis requires that the initial temperature of the model be defined in addition to already defined **Heat Power** and **Convection** coefficients (those have been copied from steady state thermal analysis).

We assume that both components have the same initial temperature of 300K. Right-click *Load/Restraint* folder in *heat sink transient* study and select **Temperature** to open window shown in Figure 8-13. Select **Initial Temperature** and enter 300K. From the fly-out menu select both assembly components.

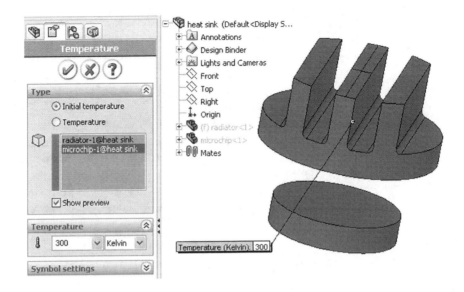

Figure 8-13: Initial Temperature specified for both assembly components

Assembly components can be selected from the SolidWorks fly-out menu.

Now run the analysis and display the temperature plot. Display the temperature plot for the last step (step number 10) by right-clicking the plot icon, selecting **Edit Definition** and setting **Plot Step** to 10 (Figure 8-14).

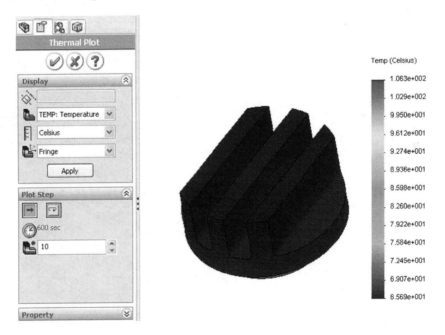

Figure 8-14: Temperature distribution 600 seconds after the heat power has been turned on.

Since we have not specified heat power as function of time, it is assumed that the full power is turned on at time t=0, when the assembly is at an initial temperature of 300K. Figure 8-14 shows the temperature distribution after 600 seconds. Notice that this result is very close to the result of steady state thermal analysis, meaning that after 600 seconds, the temperature of assembly has almost stabilized.

To see the temperature history at selected locations of the model, we proceed as follows: make sure the plot in Figure 8-14 is still showing, right-click its icon in COSMOSWorks Manager and select **Probe**. To probe temperature in the location shown in Figure 8-1, select (left-click) vertex created by the spilt lines. This probing opens the window shown in Figure 8-15.

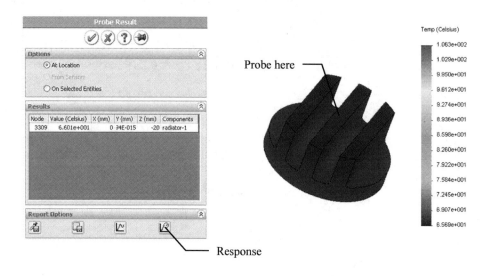

Figure 8-15: Temperature probed in the location shown in figure 8-1

Select **Response** in the **Probe** window (Figure 8-15) to display a graph showing the temperature at the probed location as a function of time (Figure 8-16).

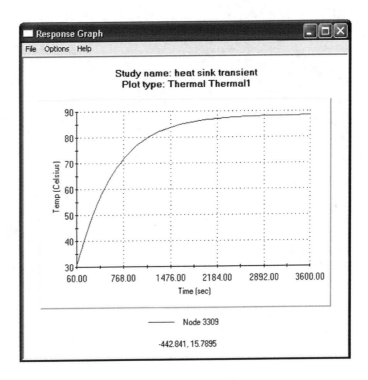

Figure 8-16: Temperature as a function of time in the probed location

To produce a response graph (here temperature as a function of time), you may probe temperature from any of the 10 performed time steps. It does not have to be the last step.

A quick examination of the graph in Figure 8-16 proves that after 3600 seconds the probed location has practically achieved the steady state temperature.

Notes:

9: Static analysis of a hanger

Topics covered

- Static analysis of assembly
- Global and local Contact/Gaps conditions

Project description

In this exercise we introduce structural analysis of assemblies. To begin, we review different options available for defining the interactions between assembly components (we began exploring this topic in the previous exercise).

Open the HANGER assembly, create a Static study, and look more closely at the pop-up menu that opens when you right-click the *Contact/Gaps* icon, already shown in Figure 8-2 and repeated in Figure 9-1 for easy reference.

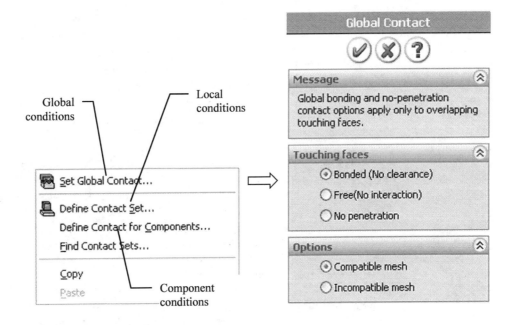

Figure 9-1: Pop-up menu associated with the *Contact/Gaps* icon

The pop-up menu, opened by right-clicking the Contact/Gaps icon folder in the COSMOSWorks Manager window, distinguishes between global, component, and local Contact/Gaps conditions. The default choice in global conditions (as well as in component and local conditions) is that touching faces are bonded.

The differences between the available Contact/Gaps conditions are as follows:

- Global contact – affects all faces in an assembly
- Contact for component – affects one component
- Local contact – affects only two specified faces (which must belong to different components of an assembly)

More detailed descriptions are given in the following tables:

GLOBAL	
Option	**Description**
Touching Faces: **Bonded**	Touching areas of different components are bonded. This option is available for structural (static, frequency, and buckling) and thermal studies.
	If touching faces are left as **Bonded** and Global conditions are not overridden by Component or Local conditions, then an assembly behaves as a part.
	This contact condition comes with two options: **Compatible mesh** and **Incompatible mesh**. Using the **Compatible mesh** option nodes located on touching faces are shared by elements on both sides. If **Compatible mesh** causes meshing difficulties, then **Incompatible mesh** option can be used. Meshes on each face are then created independently but are automatically constrained to each other.
Touching Faces: **Free**	The mesher treats parts as disjointed bodies. This option is available for structural (static, frequency, and buckling) and thermal studies. For static studies, the loads can cause interference between parts. Using this option can save solution time if the applied loads do not cause interference. For thermal studies, there is no heat flow due to conduction through touching faces.
Touching Faces: **No penetration**	The mesher creates compatible meshes on overlapping parts of touching faces. The program creates gap elements connecting the coincident nodes. This option is available for static, nonlinear, and thermal studies. For static studies, a gap element between two nodes prevents part interference, but allows the two nodes to move away from each other.

COMPONENT	
Option	**Description**
Touching Faces: **Bonded**	The mesher will bond common areas of the selected components at their interface with all other components. This option is available for structural (static, nonlinear, frequency, and buckling) and thermal studies.
	This contact condition comes with two options: **Compatible mesh** and **Incompatible mesh**. Using **Compatible mesh** option the two components are just meshed across the touching faces. Nodes located on touching faces are shared by elements on both sides. If **Compatible mesh** causes meshing difficulties then **Incompatible mesh** option can be used. Meshes on each face are then created independently but are automatically constrained to each other.
Touching Faces: **Free**	The mesher will treat the selected components as disjointed from the rest of the assembly. This option is available for structural (static, frequency, and buckling) and thermal studies. For static studies, the loads can cause interference between parts. Using this option can save solution time if the applied loads do not cause interference.
Touching Faces: **No penetration**	The mesher will create compatible meshes on overlapping areas of touching faces. The nodes associated with the two parts on the common areas are coincident but different. The program creates a gap element connecting each two coincident nodes. This option is available for static, nonlinear, and thermal studies. For static studies, a gap element between two nodes prevents part interference but allows the two nodes to move away from each other.

LOCAL	
Option	**Description**
No penetration	Available for static, drop test, and nonlinear studies only. This contact type prevents interference between source and target entities but allows gaps to form
Bonded	The source and target entities are bonded. The entities may be touching or within a small distance from each other. The program gives a warning if the distance between bonded entities is larger than the average element size of the associated elements. Only source and target entities are required to define this contact type.
Shrink fit	Valid for faces from two components which show interference. This interference is eliminated after solution.
Free	The selected faces are disjoined and can freely go through each other with no interaction.
Virtual wall	This contact type defines contact between the source entities and a virtual wall defined by a target plane. The target plane may be rigid or flexible. You can define friction between the source and the target plane.
Insulated	Available for thermal studies only. This option is similar to the Free option for structural studies. The program treats the source and target faces as disjointed. The program prevents heat flow due to conduction through the source and target entities.
Thermal resistance	Specifies thermal resistance between source and target faces.

Local conditions can be defined either with **Node to Node** option available for touching source and target faces only, or with one of two surface options: **Node to Surface** or **Surface to Surface,** which do not require compatible source and target meshes. We will now discuss the important differences between **Node to Node** and **Surface** conditions.

A **Node to Node** condition can be applied to faces that overlap. The faces do not have to be the same size; the faces just have to share some common area. **Node to Node** conditions can be specified between two:

- Flat faces
- Cylindrical faces of the same radius
- Spherical faces of the same radius

With these options, the mesh on both faces in the area where they overlap is created in such a manner that there is node to node correspondence (nodes are coincident) on both touching surfaces; hence the name **Node to Node**.

Surface contact may be specified between two faces of different shapes and can only be specified as a local condition. Initially, faces can either touch or not touch, but are expected to come in contact once the load has been applied to the model. The **Surface to Surface** condition is more general, but less numerically efficient than **Node to Surface** condition. Any contact could be defined as a **Surface** condition, but this would unnecessarily complicate the model.

The difference between **Node to Node** and **Surface** conditions is illustrated in Figure 9-2.

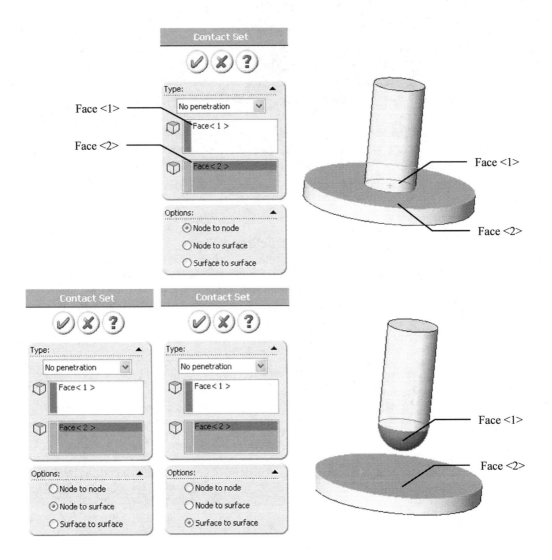

Figure 9-2: The contact between a flat end punch and a plate (top) is defined as a Node to Node contact. The contact between a spherical punch and a plate (bottom) can be defined either as Node to surface or Surface to surface contact.

Faces in Surface contact conditions don't have to touch initially. They are expected to come in contact under the load.

Global Contact/Gaps conditions can be overridden by **Component** and/or **Local** conditions. For example, using **Global** Contact/Gaps conditions, we can request that all faces be bonded. Next, we can locally override this condition and define local conditions for one or more pair as **No penetration**. The hierarchy of **Global**, **Component**, and **Local** Contact/Gaps conditions is shown in Figure 9-3.

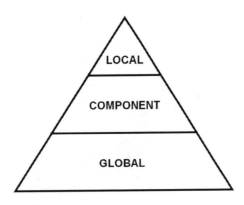

Figure 9-3: Hierarchy of Contact/Gaps conditions

In the hierarchy of Contact/Gap conditions, local conditions override both Component and Global conditions. Component conditions override Global conditions.

Procedure

Assembly HANGER (Figure 9-4) consists of three components; material (AISI 304) is assigned to all assembly components.

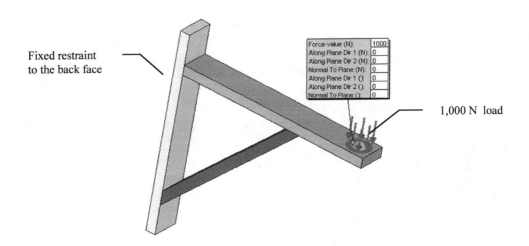

Figure 9-4: Hanger assembly

The hanger assembly consists of three parts (compare with Figure 9-6). A 1,000 N load is applied to the split face, and support is applied to the back of the vertical component.

If you do not modify the Contact/Gaps conditions, then by default all touching faces are bonded, and the assembly will behave as one part. A sample result is shown in Figure 9-5.

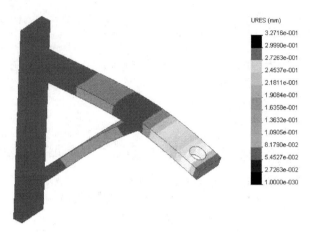

Figure 9-5: Displacement results for a model with all touching faces bonded. The assembly behaves as one part.

Notice that the hanger assembly model is adequate for analysis of displacements. However, due to stress singularities in the sharp re-entrant corners, it is not suitable for analysis stresses in these corners.

Now, we modify the Contact/Gaps conditions on selected touching faces. We leave the global conditions as **Touching Faces: bonded**, but locally we override them by defining a local **Contact/Gaps** condition. One of the three pairs of touching faces (Figure 9-6) will be defined using the local condition **Free**, meaning that there is no interaction between the faces. These faces will be able to either come apart or "penetrate" each other with no consequences.

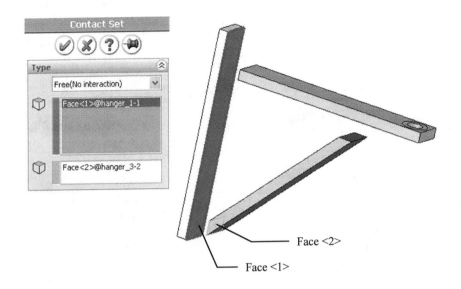

Figure 9-6: Touching Faces: Free. When faces 1 and 2 are locally defined as **Free**, there is no interaction between them.

An exploded view of the hanger makes it easier to define local Contact/Gaps conditions.

Once local or component Contact/Gaps conditions have been defined, an icon is placed in the *Contact/Gaps* folder, as shown in Figure 9-7.

Every time Contact/Gaps conditions are changed, a new mesh needs to be created (Figure 9-7).

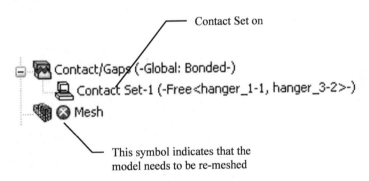

Figure 9-7: Any change in Contact/Gaps invalidates the mesh and requires remeshing

Remeshing deletes the previous results. If we wish to keep the earlier results, we need to define local Contact/Gaps conditions in a new study.

The lack of interference between faces in a contact pair locally defined as **Free** is best demonstrated by showing displacement results (Figure 9-8).

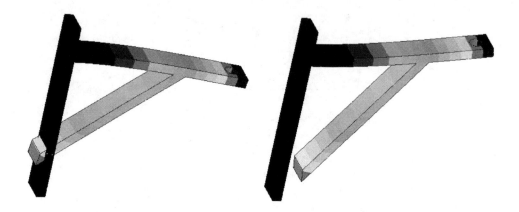

Figure 9-8: Displacement results in a pair defined as **Free**

The plot on the left shows results for load directed downwards and the plot on the right has the load direction reversed.

Let's now change the local contact conditions between the faces shown in Figure 9-6 from **Free** to **No penetration** using the **Node to node** option. Remesh the model, and run the solution again. Notice that the solution now requires much more time to run because the contact constraints must be resolved (Figure 9-9). Any contact conditions represent nonlinear problems and require an iterative solution.

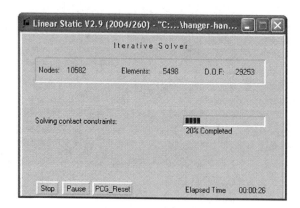

Figure 9-9: Iterative solution for the hanger assembly using a No penetration condition

A No penetration condition requires an iterative solution to solve contact constraints and takes significantly longer to complete than a linear solution.

The displacement results, displayed in Figure 9-10, show that the two faces defined as **No Penetration** now slide when a downward load is applied (left) and separate when the load is applied upward (right).

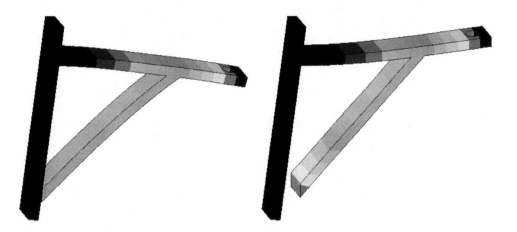

Figure 9-10: Displacement results for two locally defined Node to Node faces

The two faces in the pair defined locally as Node to Node slide (left) or separate (right) depending on the load direction.

Closer examination of the displacement results for the sliding faces (Figure 9-11) shows that the sliding faces partially separate.

Figure 9-11: Partial separation of the sliding faces

Only a portion of face 1 contacts face 2. The mesh in the contact area is too coarse to allow for the analysis of contact stresses.

While the mesh is adequate for the analysis of displacements, the mesh is not sufficiently refined for the analysis of the contact stresses that develop between the two sliding faces.

Notes:

10: Analysis of contact stresses between two plates

Topics covered

- Assembly analysis with surface contact conditions
- Contact stress analysis
- Avoiding rigid body modes

Project description

We will perform an analysis of contact between two plates in assembly model TWO PLATES. The model consists of two identical plates that touch each other on their curved outside surface (Figure 10-1). The material for both parts is Nylon 6/10 and has already been assigned to the parts. Our objective is to find the distribution of von Mises stress and the maximum contact stresses under a 1,000N of compressive load.

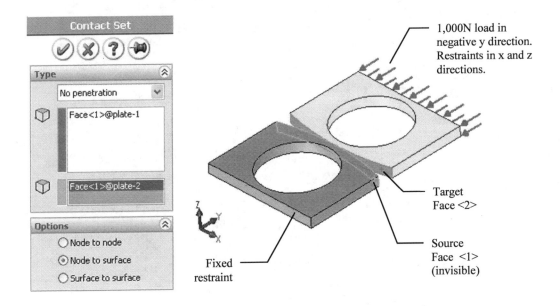

Figure 10-1: Two plates with their cylindrical surfaces in contact
It does not matter which face is the source and which is the target.

Preparation of the model for analysis requires restraining the loaded part to prevent rigid body motion and at the same time, make it free to move in the direction of the load. This can be accomplished by restraining the loaded face in both in-plane directions while leaving the normal direction (the direction in which load is applied) unrestrained (Figure 10-2).

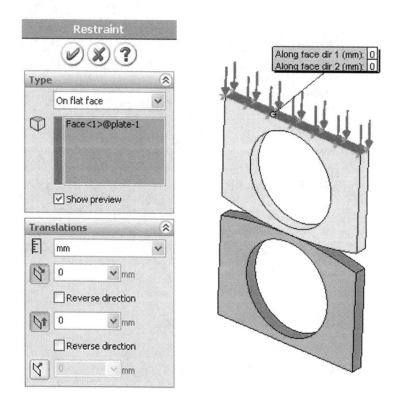

Figure 10-2: Restraint windows

Restraints are conveniently applied to the loaded face defining 0 prescribed displacements in both in-plane directions of the face.

The restraints shown in Figure 10-2 are required to prevent rigid body motions: rolling motion, and sliding the top plate off the bottom plate. The analyzed contact problem is frictionless.

We are now ready to mesh the assembly model. Adequate mesh density in the contact area is of paramount importance in any contact stress analysis. It is the responsibility of the user to make sure that there are enough elements in the contact area to properly model the distribution of contact stresses. In this exercise we use a default global element size and apply mesh controls to contacting faces as shown in Figure 10-3.

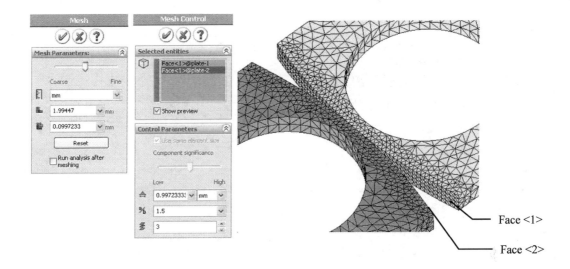

Figure 10-3. Global mesh size and mesh control (applied to contacting faces) used to create the mesh. Global element size is 2mm and mesh control size is 1mm.

Do not use Automatic transition.

Shown in Figure 10-4 is a contact pressure plot and associated windows used to define it.

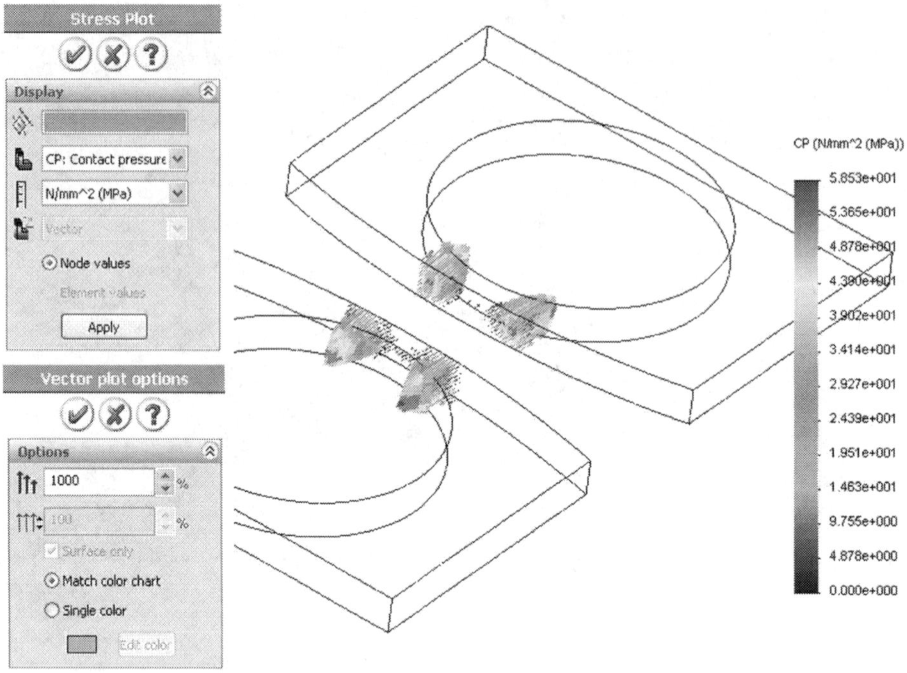

Figure 10-4: Contact stress results for the coarse mesh presented using an exploded view. The maximum contact stress reported is 59 MPa.

The plot presents Contact pressure defined in the Stress Plot window (left top). Contact pressure plot uses a Vector type of display which can be modified using Vector plot options window (left bottom).

Von Mises stresses are shown in Figure 10-5.

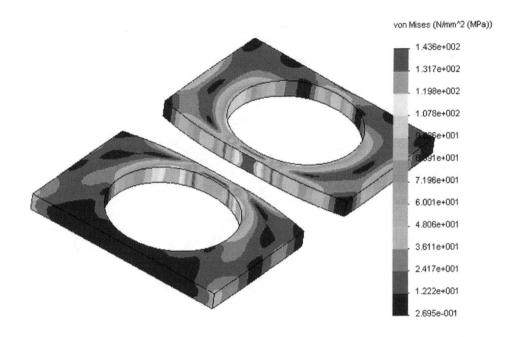

Figure 10-5: Von Mises stress results presented using exploded view

We leave it to the reader to decide if this stress level (the maximum von Mises stress in contact area is 144MPa) is acceptable for Nylon 6/10 material which has yield stress of 139 MPa.

Before concluding this exercise, further investigate the effects of mesh refinement and type of material (such as steel or aluminum) on contact stresses. Note that our mesh is adequate for modeling contact stresses between two Nylon parts because low modulus of elasticity of Nylon makes the contact area quite large. The same mesh may prove too coarse when analyzing more rigid materials.

Notes:

11: Thermal stress analysis of a bi-metal beam

Topics covered

- Thermal stress analysis
- Use of soft springs to stabilize the model
- Shear stress analysis

Project description

The temperature of the entire bi-metal beam, shown in Figure 11-1, increases uniformly from initial 298K to 600K. We need to find thermal stresses induced by that increase in temperature.

Procedure

Open the assembly file BIMETAL. Note that the assembly consists of two instances of the same parts. For this reason material can not be assigned to the SolidWorks part. We must assign material properties in COSMOSWorks.

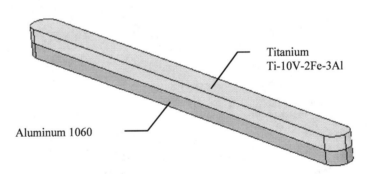

Figure 11-1: Bi-metal beam consisting of bonded titanium and aluminum strips

The bi-metal beam will deform when heated because of the different thermal expansion ratios of titanium and aluminum.

To account for thermal effects we define the study as **Static,** and in the **Properties** of this study, under **Flow/Thermal Effects** select the option **Input temperature** (Figure 11-2).

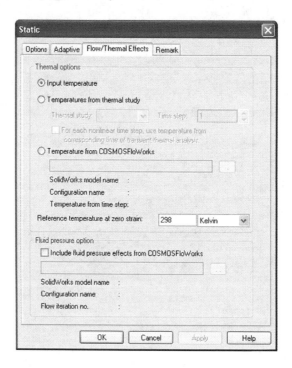

Figure 11-2: Flow/Thermal Effect tab in Static study window, where we instruct the solver to account for the effects of Input temperature and define the reference temperature at zero strain, which we define here as 298K

Before proceeding, let's take this opportunity to review all thermal options available in the study window.

Thermal Option	Definition
Input temperature	Use if prescribed temperatures will be defined in the *Load/Restraint* folder of the study to calculate thermal stresses. This is our case.
Temperature from thermal study	Use if temperature results are available from a previously conducted thermal study.
Temperature from COSMOSFloWorks	Use if temperature results are available from a previously conducted COSMOSFloWorks study.

Now assign the material properties of titanium and aluminum to the assembly components as shown in Figure 11-1.

There are no restraints in this model, which approximates the situation of the bimetal resting on a table, or more exactly, floating in space in the absence of gravity. The unsupported model has six rigid body motions which can not be tolerated in structural analysis. Therefore, we need to add a very low, artificial stiffness to enable the solution. For that we must check Check the **Use soft springs to stabilize model** solver option on the **Options** tab of the static study (Figure 11-3).

Alternatively, **Use inertial relief** solver option could be used. In this case, the model would be stabilized (rigid body modes would be eliminated) by applying a very low inertial load to counteract any load imbalance arising from numerical errors.

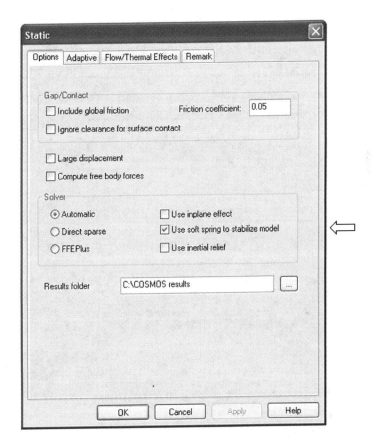

Figure 11-3: Check *Use soft springs top stabilize model* option to remove rigid body motions

To apply temperature load, right-click the *Load/Restraint* folder and select **Temperature** to open the **Temperature** window. From the fly-out menu select both assembly components and enter a temperature of 600K, which means that assembly temperature will be increased by 302K from the initial 298K (Figure 11-4).

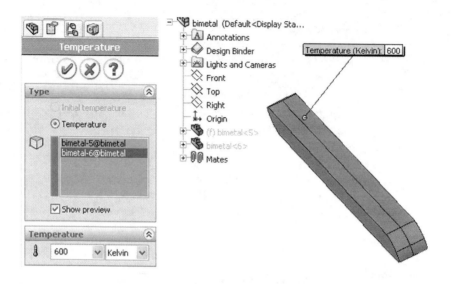

Figure 11-4: A temperature of 600K is applied to both assembly components, which are most conveniently selected from the SolidWorks fly-out menu

Mesh the model with a 1mm element size to create a total of six layers of elements across the model's thickness.

Displacement results are shown in y 11-5, von Mises stress results are shown in Figure 11-6.

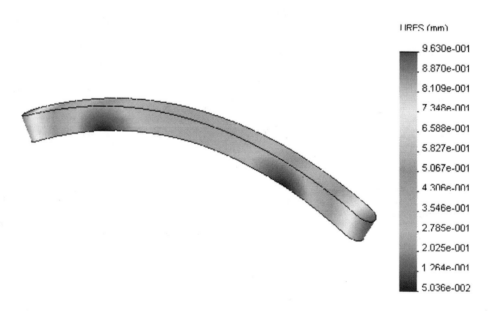

Figure 11-5: Displacement results for the bi-metal beam

Due to the different thermal expansion ratios of titanium and aluminum, thermal strains develop and bend the bimetal beam.

Note that in absence of supports, displacement results are difficult to interpret; there is no reference system to measure them from. The Plot shows displacements relative to the center of mass of the model.

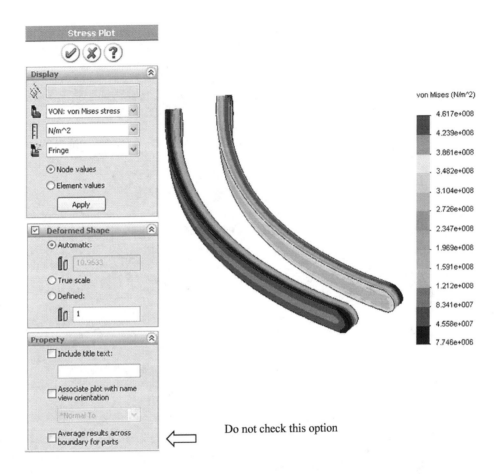

Figure 11-6: Von Mises stress results for the bi-metal beam. Make sure that the option Average results across boundary for parts is not checked.

It may be interesting to review shear stresses on the boundary between these two parts. Considering assembly orientation in the global coordinate system, we need to display the TZY and TZX shear stress components selectable in the **Stress Plot** window as shown in Figure 11-7.

Engineering Analysis with COSMOSWorks Professional 2007

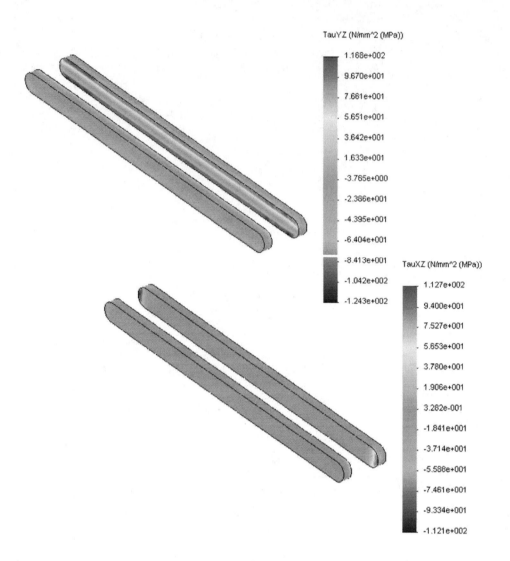

Figure 11-7: Two shear stress components on the face joining two beams

Model is shown in exploded view. Consult Figure 1-11 and Figure 11-8 for more information on the stress cube and symmetry of shear stresses.

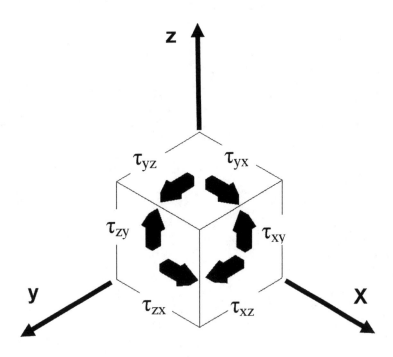

Figure 11-8: Six shear stress components

Note that the equilibrium condition requires that $\tau_{xy} = \tau_{yx}$, $\tau_{yz} = \tau_{zy}$, $\tau_{xz} = \tau_{zx}$

The following convention is used in COSMOS to determine directions of shear stresses:

TXY: Shear in Y direction on YZ plane

TXZ: Shear in Z direction on YZ plane

TYZ: Shear in Z direction on XZ plane

Notes:

12: Buckling analysis of an L-beam

Topics covered

- Buckling analysis
- Buckling load safety factor
- Stress safety factor

Project description

An L-shaped perforated beam is compressed with a 40,000 N load, as shown in Figure 12-1. The material has already been assigned to the assembly components. Our goal is to calculate the factor of safety related to the yield stress and the factor of safety related to buckling.

Procedure

Open the assembly file L BEAM. The beam and endplates material is alloy steel with yield strength of 620 MPa.

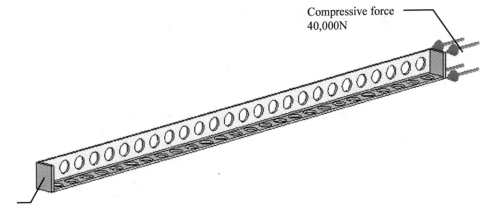

Figure 12-1: L-beam assembly model

A perforated angle is compressed by a 40,000 N force uniformly distributed over the endplate.

Before we run a buckling analysis, let's first obtain the results of a static analysis based on the load and restraint shown in Figure 12-1. Name this study *stress analysis*. The results of static analysis show the maximum stress below the yield strength 620 MPa. Figure 12-2 identifies the location of the highest stress.

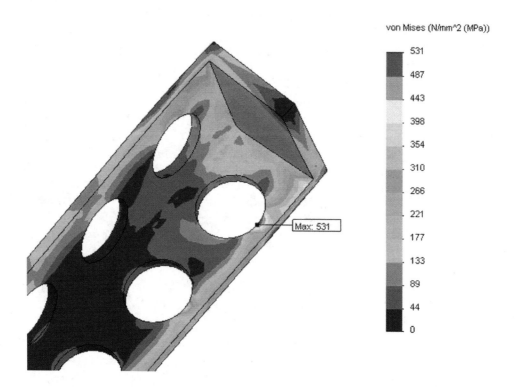

Figure 12-2: Location of the maximum von Mises stress is close to the loaded endplate

Plot displays the location of the maximum stress and uses floating format to display numerical values. Both are selected in Chart Options.

As is always the case with slender members under a compressive load, the factor of safety related to material yield stress may not be sufficient to describe the structure's safety. This is because of the possible occurrence of buckling. We need to calculate the buckling safety factor which requires running a buckling study created as shown in Figure 12-3.

Copy loads and restraints from the *stress analysis* study to *buckling analysis* study.

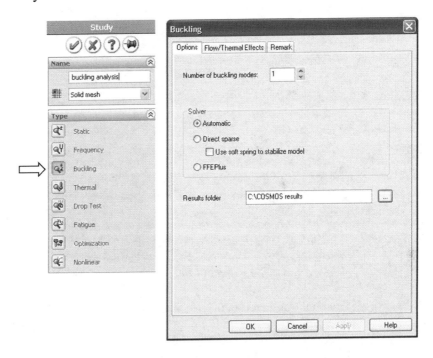

Figure 12-3: Definition of a buckling study (left) and the Buckling study window, where properties are defined for a buckling study.

Defining a buckling study requires specifying the number of desired buckling modes. Here we ask for first buckling mode.

When defining a buckling study, we need to decide how many buckling modes should be calculated. This is a close analogy to the number of modes in a frequency analysis. In most practical cases, the first buckling mode determines the safety of the analyzed structure. Therefore, we limit this analysis to calculating the first buckling mode. Once the buckling analysis has been completed, COSMOSWorks automatically creates two result folders: *Displacement* and *Deformation*.

Even though displacement results are available, they do not provide any useful information. In a buckling analysis, the magnitude of displacement is meaningless, just like in frequency analysis. The deformation plots shown in Figure 12-4 do not include confusing information on the magnitude of displacement; instead it lists the buckling load factor.

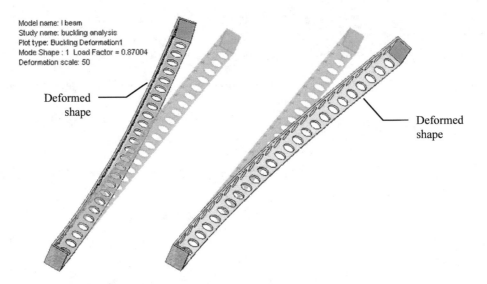

Figure 12-4: Deformation plot provides visual information on the shape of the buckled structure. It also lists the buckling factor, here equal to 0.87. This plot shows the buckled shape along with the undeformed model.

Buckling may take place in either direction. Shown on the left is the direction that COSMOS assigned a positive value on the scale of deformation, and on the right the direction that COSMOS assigned a negative value on the scale of deformation. This can be changed manually in the Deformation plot definition.

The buckling load factor provides information on how many times the load magnitude would need to be increased in order for buckling to actually take place. In our case, the magnitude of load causing buckling is 0.87*40,000N = 34,800 N. Therefore, buckling will take place because the actual load is 40,000N. The buckling load factor can be also called the buckling load safety factor. Notice that the calculated buckling load safety factor is actually lower than the previously calculated safety factor related to material yield strength. Therefore the beam will buckle before it develops stresses exceeding the yield strength.

Our conclusion is that buckling is the deciding mode of failure. Also notice that high stress affects the beam only locally, while buckling is global.

It should be pointed out that the calculated value of the buckling load is non-conservative, meaning that it does not account for the always-present imperfections in model geometry, materials, loads, and supports. Also, it does not account for the fact that the meshed model is stiffer than the corresponding model before meshing. With this in mind, the real buckling load may be significantly lower than the calculated 34,800N.

13: Design optimization of a plate in tension

Topics covered

- Structural optimization analysis
- Optimization goal
- Optimization constraints
- Design variables

Project description

A rectangular plate with two holes is subjected to a 10,000 N tensile load resisted by a fixed restraint (Figure 13-1). We suspect that the plate is over-designed and wish to find out if material can be saved by enlarging the diameter of the holes. Because of certain design considerations, the hole diameter cannot exceed 40 mm. From previous experience with similar structures, we know that the highest von Mises stress should not exceed 300 MPa anywhere in the plate.

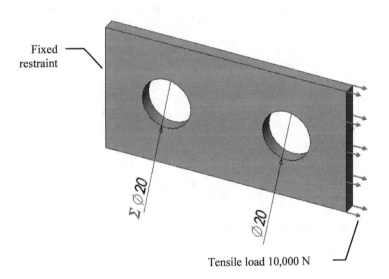

Figure 13-1: Rectangular plate with two round holes is subjected to a 10,000 N tensile load

The model is shown in its initial configuration, before optimization. Diameter of both holes is controlled by the hole dimension on the load side (review equation in SolidWorks model).

As with every design optimization problem, this one is defined by the optimization goal, design variables, and constraints. These terms are explained below.

Term	Definition
Optimization goal	Our objective is to minimize the mass. Other examples of optimization goal are to maximize stiffness, maximize the first natural frequency, etc. The optimization goal is also called the optimization objective or optimization criterion.
Design variable	The entity that we wish to modify is a design variable. In this example, the design variables are the hole diameters. The range for the diameter is from the initial 20mm to a maximum of 40mm.
Constraints	Limits on the maximum deflection, maximum stress or minimum natural frequency are examples of constraints. Constraints in optimization exercises are also called limits. In this exercise, the constraint is the maximum von Mises stress, which must not exceed 300 MPa.

Procedure

Open and review part file EC022. It comes with material defined as Alloy steel.

Before starting the design optimization exercise that will result in changing the diameters of the holes, let's find the stresses in the plate "as is," which requires running a static study. We would have to set up **Static** study anyway because **Static** study is a pre-requisite to the subsequent optimization study. Information on material, loads and restraints required by the optimization study are taken from that prerequisite **Static** study.

Von Mises stress results in plate before optimization are shown in Figure 13-2.

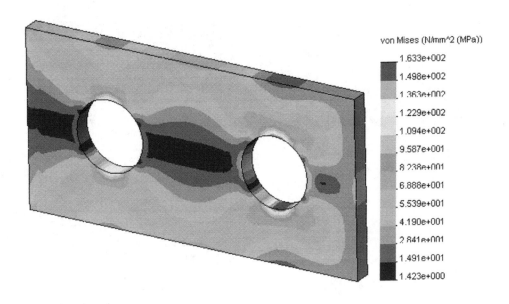

Figure 13-2: Von Mises stresses in the model before optimization

Von Mises stress results show the maximum von Mises stress of 163 MPa at the edge of the hole closer to the load. This is below the allowable 300 MPa, so the plate is indeed over designed and we can proceed with the optimization exercise.

The optimization study is defined as any other study except that mesh does not need to be defined since the optimization study must use the same type of mesh as the prerequisite static study (Figure 13-3).

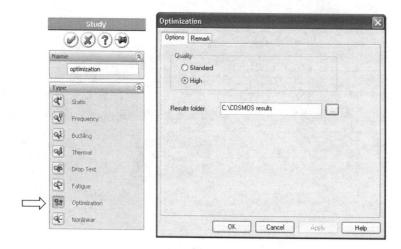

Figure 13-3: Optimization study definition and optimization study properties

High quality is selected on the Options tab. The Standard quality option uses a smaller number of iterations to determine the optimum design.

When the optimization study is defined, COSMOSWorks automatically creates three folders specific to an optimization study (Figure 13-4):

❑ *Objective*

❑ *Design Variables*

❑ Constraints

All three folders need to be edited prior to running the optimization analysis.

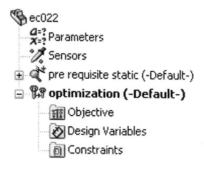

Figure 13-4: Design optimization study contains three automatically created folders: Objective, Design Variables, and Constraints.

Note a static study must be defined prior to defining optimization study. Here, we named this static study: pre requisite static.

The optimization goal is defined in the **Objective** window. To edit this window, right-click the *Objective* folder (Figure 13-5). For this exercise, we accept the default optimization goal, which is to minimize the mass.

Figure 13-5: Objective definition window

Note that a static study that precedes the optimization study is selected when Objective is defined. If more than one preceding study exists, you must choose one of them to associate with the Optimization study.

Make sure the hole dimensions are visible. You can display the dimensions by selecting **Show Feature Dimensions** in the *Annotations* folder in SolidWorks Manager.

Right-click the *Design Variables* folder and select **Add** from the pop up menu to open the **Design Variable** window. Select the dimension of the hole on the load side (the independent variable in equation) and enter the allowed range of diameter change (20mm – 40mm) as shown in Figure 13-6. Repeat the same to define the second design variable controlling the other hole.

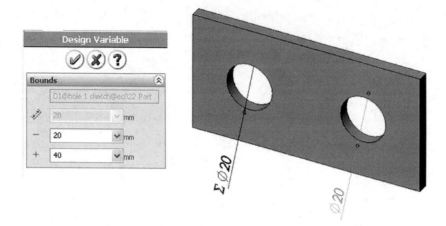

Figure 13-6: Definition of design variable

The allowed range of the hole diameter is specified as from 20 mm (value before optimization) to 40 mm, which is the maximum allowed hole diameter.

The diameter of the right hole is a design variable. Although an equation in the SolidWorks model, it also controls the diameter of left hole (both hole diameters are equal).

Note that even though the allowed variation of the hole diameter is from 20 mm to 40 mm, it may not be possible to reach the diameter of 40 mm if during the process of increasing the hole diameter the maximum stress exceeds 300 MPa.

To define the constraints, right-click the *Constraints* folder to open the **Constraint** window, shown in Figure 13-7.

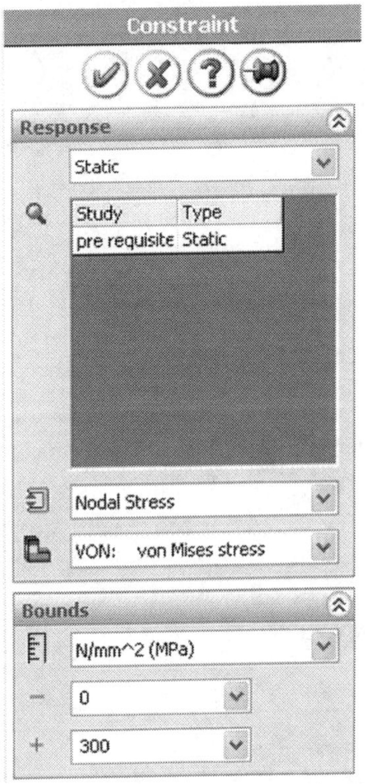

Figure 13-7: Constraint window

The maximum allowed stress is defined as nodal von Mises stress equal to 300 MPa.

Notice that loads, restraints, and materials are not defined anywhere in the optimization study. The necessary information is transferred from the prerequisite static study specified in the **Objective** and **Constraint** window (Figures 13-5 and 13-7).

Having defined the optimization goal, design variable(s), and constraint(s) (Figure 13-8 left), we are ready to run the optimization study. When the solution is complete, COSMOSWorks creates several result folders, which are shown in Figure 13-8 (right).

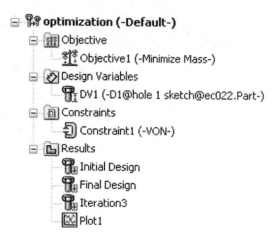

Figure 13-8: Fully defined folders: Objective, Design Variables, Constraints (left) and Result folders

To view the optimized model, double-click the *Final Design* icon in the *Design Cycle Result* folder. To view the original model for comparison, double-click the *Initial Design* icon in the *Design Cycle Result* folder (Figure 13-9).

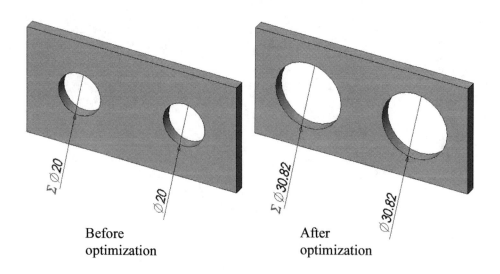

Figure 13-9: Model before optimization (left) and after optimization (right)

In the optimized model, the diameter of the holes is 30.82 mm.

If desired, we can display the model configuration in any step of the iterative design optimization process. To open the **Design Cycle Result** window, right-click the *Design Cycle Result* folder and select the desired iteration number (Figure 13-10).

Figure 13-10: Design Cycle Result window

Use it to display model configuration after the specified iteration.

To examine displacements or stresses in the optimized model, we need to review the plots in the prerequisite static study, the results of which have been updated to account for the optimized model geometry.

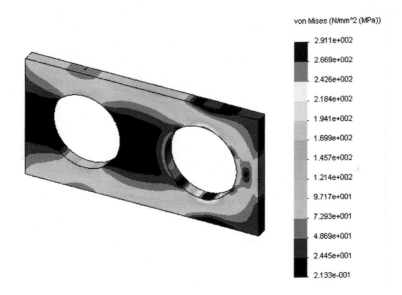

Figure 13-11: Von Mises stresses in the optimized model

Note that the maximum stress is 300 MPa which is the maximum allowed stress as defined in Constraint window.

Analyzing results presented in Figure 13-11 it is clear that optimization process stopped because the maximum stress level was reached, not because the maximum allowed hole diameter was reached.

The history of the optimization process can be reviewed by examining plots in the *Design History Graph* folder and in the *Design Local Trend Graph* folder. An example is shown in Figure 13-12.

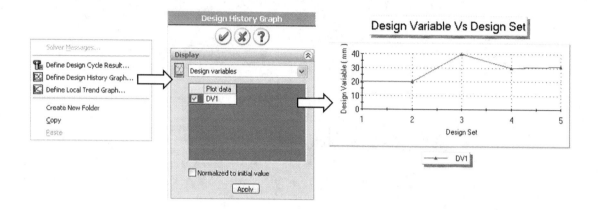

Figure 13-12: Graph showing changes in design variable during the iterative optimization process.

To open the Design History Graph, right-click the Results folder, then select the desired graph, here the Design History Graph.

14: Static analysis of a bracket using adaptive solution methods

Topics covered

- Comparison between h-elements and p-elements
- h-Adaptive solution method
- p-Adaptive solution method

Project description

A bracket shown in Figure 14-1, is supported along the backside. A bending load of 10,000 N is uniformly distributed on the cylindrical face. We need to find the location and magnitude of the maximum von Mises stress. Open the part file BRACKET. The part material is AISI 304.

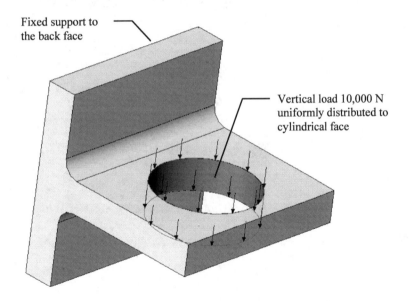

Figure 14-1: Hollow cantilever bracket under a bending load

Due to the symmetry of the bracket geometry, loads, and supports, we could simplify the geometry further by cutting it in half, but decide against it because the work involved would not save time overall.

We use this simple problem to introduce two new solution methods: the h-adaptive method using h elements, and the p-adaptive solution method using p elements. h elements are what we have been using up until now, so before we proceeding, an explanation of p-elements will be given.

If you recall, in Chapter 1 we mentioned that COSMOSWorks can use either first order elements (called draft quality), or second order elements (called high quality). We also said that first order elements should be avoided.

Furthermore, recall that first order elements model a linear (or first order) displacement distribution and constant stress distribution, while second order elements model a parabolic (second order) displacement distribution and linear stress distribution. We now have to amend the above statements.

Besides first and second order elements, COSMOSWorks can work with elements up to the fifth order. These higher order elements (called p-elements) can be accessed if **p-adaptive** is selected as the solution method in the Study window under the **Adaptive** tab (Figure 14-2). This option is available only for static analysis using solid elements.

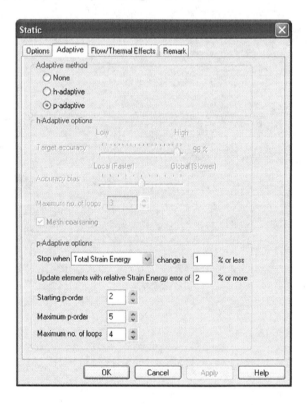

Figure 14-2: Adaptive tab in the Static window with p-adaptive solution method selected

The selection "p-adaptive" made on the Adaptive tab in the Static study window activates the use of the p-adaptive solution method. For the p-adaptive solution presented in this chapter, we use the settings as shown in this illustration.

p-adaptive solution options shown in Figure 14-2 are explained in table below:

Setting	Definition
Stop when	Looping continues until the change in Total Strain Energy (or other measure like RMS resultant displacement or RMS von Mises stress) between the two consecutive iterations is less than 1 percent, as specified in the p-adaptive options area. If this requirement is not satisfied, then looping will stop when the elements reach the fifth order; this will be fourth loop.
Update elements with relative Strain Energy error of ...	This setting controls which elements are upgraded during the iterative solution
Starting p-order	Initial order of elements, usually the starting p-order, is set to 2, which means that all elements are first defined as second order elements.
Maximum p-order	The actual highest order to be used in a p-adaptive solution. The highest order available in COSMOSWorks is the fifth order.
Maximum no. of loops	Sets the maximum number of loops allowed when you run the study. The maximum possible number of loops is four.

Elements used in p-Adaptive solutions do not have a fixed order, and can be upgraded "on the fly" (automatically) during the iterative solution process without our intervention. These types of elements with upgradeable order are called p-elements.

Let's pause for a moment and explain some terminology.

Refer to Figure 2-16 which explains that h denotes the characteristic element size. While the mesh is refined during the convergence process, the characteristic element size, h, becomes smaller. Therefore the mesh refinement process that we conducted in Chapters 2 and 3 is called the h-convergence process, and the elements used in this process are called h-elements. Note that h-elements retain their order; they cannot be upgraded to a higher order.

The iterative process that we are discussing now does not involve mesh refinement. While the mesh remains unchanged, the element order changes from the second all the way to fifth, or less if the convergence criterion (here the change in **Total Strain Energy**) is satisfied sooner.

The element order is defined by the order of polynomial functions that describe the displacements in the element. Because the polynomial order experiences changes, the process is called a p-convergence process, and the upgradeable elements used in this process are called p-elements.

Not all p-elements are upgraded during the solution process. Which elements are updated depends on the selection made in the field **Update elements with relative Strain Energy error of ___ % or more**. Here we set it to 2, meaning that only those elements not satisfying the above criterion will be upgraded (investigate other criteria as well). Therefore we say that element upgrading is "Adaptive", or driven by the results of consecutive iterations.

The p-adaptive solution process is analogous to the process of mesh refinement, which also continues until the change in the selected result is no longer significant (refer to Chapter 2). The multi step mesh refinement process can also be automated, which brings us to the second adaptive method we will study in this chapter: h-adaptive (Figure 14-3).

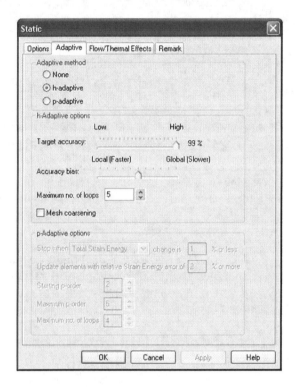

Figure 14-3: Adaptive tab in the Static window with h-adaptive solution method selected.

The selection "h-adaptive" made in the Adaptive tab in the Static study window activates the use of h-adaptive solution method. For the h-adaptive solution presented in this chapter, we use the settings as shown in this illustration.

h-adaptive solution options are explained in table below.

Setting	Definition
Target accuracy	Sets the accuracy level for the strain energy norm. This is NOT the stress accuracy level. However, a high level of accuracy in the convergence of the strain energy norm indicates good stress results.
Accuracy bias	You can move the slider towards Local to instruct the program to concentrate on getting accurate peak stress results using a fewer number of elements. Or you can move the slider towards Global to instruct the program to concentrate on getting overall accurate results.
Maximum no. of loops	Sets the maximum number of loops allowed when you run the study. The maximum possible number of loops is five.
Mesh coarsening	Check this option to allow the program to coarsen the mesh in regions with low error during the adaptive loops. The number of elements in consecutive loops may increase or decrease depending on the model and the initial mesh. If this option is not checked, the program does not change the mesh in regions with low errors. The number of elements in this case keeps increasing in each adaptive loop.

Procedure

We will solve the same problem three different ways:

1. Using one mesh of h elements (standard approach)

2. Using h-adaptive solution method

3. Using p-adaptive solution method

First solve the model using second order solid tetrahedral h-elements, and name the study *standard*. Use the default element size, but to better capture stresses, apply **Mesh Control** to both fillets (Figure 14-4) with **Automatic transition** in the **Options** window under the **Mesh** tab, even though **Automatic transition** and **Mesh Control** are seldom combined in one mesh.

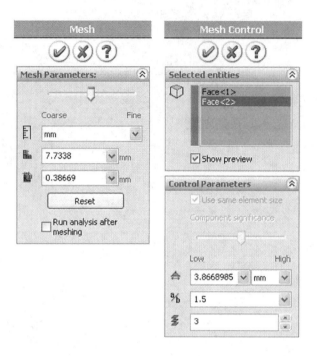

Figure 14-4: Mesh window (left) and Mesh Control window (right) used to create an h-element mesh

Figure 14-5 shows von Mises stress results obtained from the standard study. It also shows the standard h element mesh.

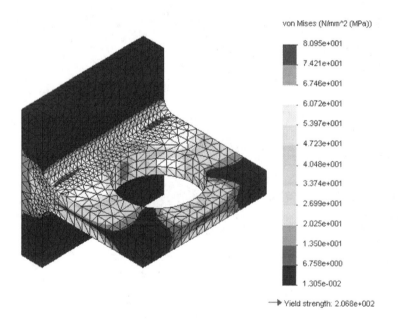

Figure 14-5: Von Mises stress results obtained using standard h-elements.

The maximum von Mises stress is 81 MPa.

Now, create a new study *h adaptive* identical the one we just finished, except that in the study window under the **Adaptive** tab, select **h-adaptive**. Use the following settings: **Target accuracy 99%**, no **Accuracy bias** (slider in the middle), **Maximum no. of loops 5**. Restraints and Loads can be copied from the previous study we just completed.

As we have already explained, the finite element mesh is refined during the h-adaptive solution and results are reported for the last, most refined mesh. To better visualize the h adaptive solution method, we start with a coarse mesh as shown in Figure 14-6. This way the automatic mesh refinement will be easier to see.

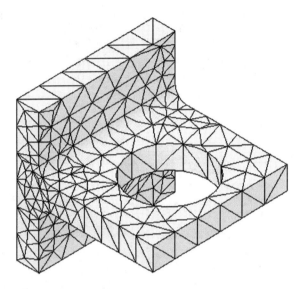

Figure 14-6: The initial mesh in *h adaptive* study

Use 20mm for element size to create this coarse mesh without any bias. If we were to run a non-adaptive solution, this mesh would be inadequate to capture stress field along the fillets.

Having created the initial coarse mesh, run the solution of the h-adaptive study and display the results as shown in Figure 14-7.

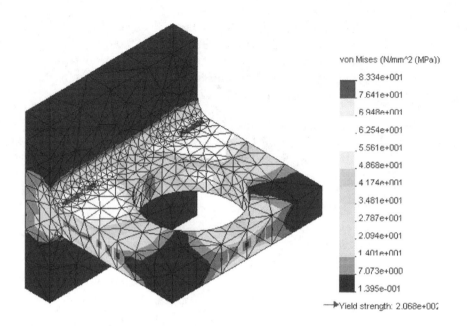

Figure 14-7: Von Mises stress results obtained with h-adaptive solution

Notice that mesh has been refined as compared to the initial mesh in Figure 14-7. The maximum von Mises stress is 83MPa.

To complete the exercise, create one more study called *p adaptive*. In the study window under the **Adaptive** tab, select **p-adaptive**. Use all defaults for p-adaptive study definition as shown in Figure 14-2. **Restraints** and **Loads** can be copied from any of the two previous studies.

Considering that the p-adaptive solution will be used, we can again manage with a coarse mesh. Create a mesh with default element size 20mm without **Automatic transition**. Using higher order elements, which is equivalent to refinement of an h-element mesh, even this coarse mesh will deliver accurate results. Indeed, having solved the study with p-elements, we produce the stress plot shown in Figure 14-8. The same figure also shows the stress plot produced by a non-adaptive solution using the same mesh.

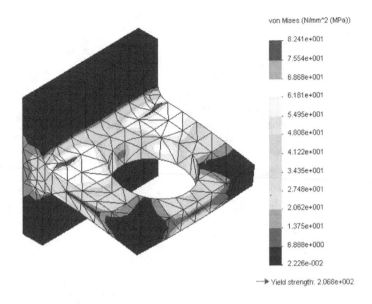

Figure 14-8: Von Mises stress results obtained using p-adaptive

The maximum von Mises stress is 82 MPa

To summarize results, we create convergence graphs which are available for both adaptive solutions (either *h* or *p)*.

To create a convergence graph, right-click the *Results* folder and select **Define Adaptive Convergence Graph**. For both studies select **Maximum von Mises Stress** in the **Convergence Graph** window (Figure 14-9). Having examined convergence of von Mises stress, try experimenting with other **Convergence Graph** options.

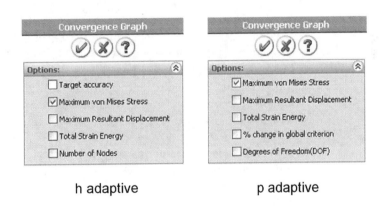

Figure 14-10: Convergence criteria available for h-adaptive and p-adaptive solutions.

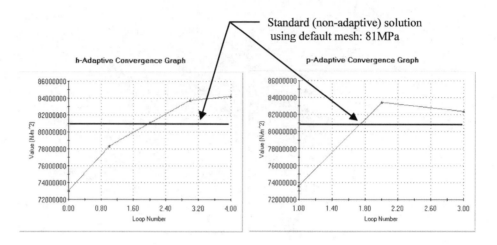

Figure 14-11: Convergence graph of Von Mises stress results obtained in h-adaptive (left) and p-adaptive (right) solutions

These graphs also show that five iterations were required to converge within the specified accuracy for h adaptive and three for p adaptive solutions. Consult Figure 15-10 for options on modifying graph properties. The graphs here have been modified to have the same Y axis scale.

Note that while the h convergence process was controlled by **Target accuracy**, p-convergence process was controlled by **Total Strain Energy**, You are encouraged to repeat both h and p convergence studies using a more refined mesh and different convergence criteria.

Which of the three solution methods are better: the "regular" method using h-elements (here the *standard* study), the h-adaptive solution (*h adaptive* study) or the p-adaptive solution (*p adaptive* study)?

Experience indicates that second order h-elements offer the best combination of accuracy and computational simplicity. For this reason, the automesher is tuned to meet the requirements of an h-element mesh intended for standard solution with h-elements. Both adaptive solution methods are much more computationally demanding and significantly more time-consuming. Therefore, the adaptive solution methods are reserved for special cases where, the solution accuracy must be known explicitly.

Both h and p adaptive methods are great learning tools, leading to better understanding of element order, the convergence process, and discretization error. For this reason, readers are encouraged to repeat some of previous exercises using adaptive solution methods.

15: Design sensitivity analysis of a hinge supported beam

Topics covered

❑ Design sensitivity analysis using Design Scenario

Project description

A beam is loaded with a 1,000N uniformly distributed load (500N to each split face) and supported by two pins placed in lugs, which are initially spaced out at 200mm (Figure 15-1).

We wish to investigate the beam deflection at locations 1 and 2 (Figure 15-1), while the distance between the lugs changes from 60mm to 340mm (Figure 15-2). Open part file BEAM WITH LUGS. The model has material 1060 Alloy already assigned.

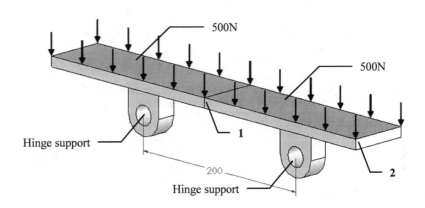

Figure 15-1: Beam with lugs in the initial configuration

The numbered vertexes on the beam are where we need to determine deflection, while the position of the lugs is changed. The range of distance between lugs is shown in Figure 15-2.

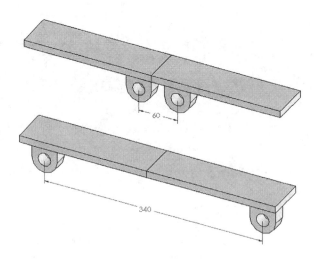

Figure 15-2: Model shown in two extreme configurations

The distance between two lugs is controlled by the dimension in Sketch 2 in the SolidWorks model.

The pins themselves are not modeled; instead the pin support is modeled by restraints applied to cylindrical faces of both lugs.

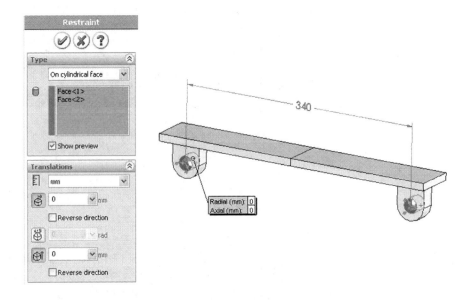

Figure 15-3: The presence of pins, which provide hinge support for the lugs, is modeled by using the On cylindrical face restraint type. Only circumferential translations are allowed on cylindrical faces of both lugs.

Hinge restraint could have been used as well.

Procedure

Define loads and restraints as shown in Figure 15-1.

If we proceeded by changing the distance from 60 mm to 340 mm in 40-mm intervals, this would require running eight analyses and a rather tedious compiling of all results. COSMOSWorks offers an easier way to accomplish our objective by implementing **Design Scenario**. Using **Design Scenario**, the distance defining the position of the holes is defined as a **Parameter** and is automatically changed in desired intervals. This allows for an automated analysis of all configurations. The results of the design scenario can then be plotted using COSMOSWorks tools.

A **Design Scenario** is often called a sensitivity study, as it investigates the sensitivity of the selected system responses (here beam deflection) to changes in certain parameters defining the model (here the distance between two lugs).

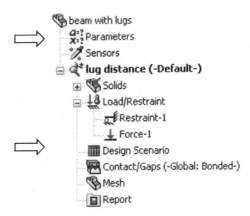

Figure 15-4: Parameters and Design Scenario folders

Parameters and Design Scenario folders are created automatically but are used only when a Design Scenario is run. Notice that the Parameters folder is created before any study is defined.

Right-click the *Parameters* folder and select **Edit/Define...** to open the **Parameters** window (Figure 15-5 top). In the **Parameters** window, select **Add**, which opens the **Add Parameters** window (Figure 15-5, bottom). In the Filter option, select **Model dimensions.** We want to select the dimension that will be undergoing changes. Doing this requires that the model dimensions are visible. The easiest way to display the dimensions is to right-click the *Annotations* folder in the SolidWorks Manager window. This opens the associated pop-up menu, from which select **Show Feature Dimensions**. With dimensions showing we can select the desired dimension.

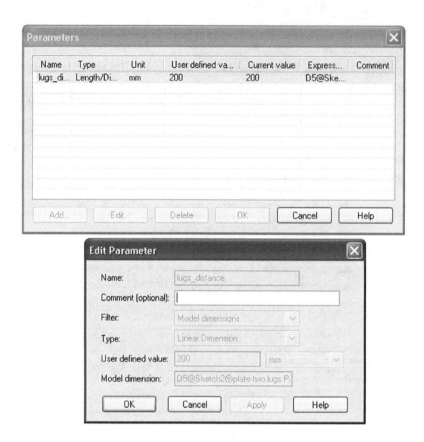

Figure 15-5: Parameters window (top) and Edit Parameter window (bottom)

In this exercise, lug_distance is the parameter defining the distance between two lugs.

Having defined the parameter, we can now define the **Design Scenario**. Right-click the *Design Scenario* folder to open a pop-up menu and select **Edit/define**. Since we wish to change the distance between the lugs from 60 mm to 340 mm in 40-mm intervals, the number of scenarios is eight. Because more than one parameter can be defined in the **Define Scenarios** tab, the parameter must be checked as active (Figure 15-6).

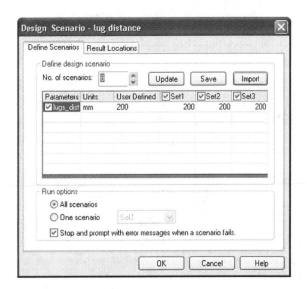

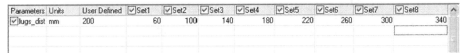

Figure 15-6: Define Scenarios tab in the Design Scenario window

To expand the table to 8 columns, click Update button. Each one of 8 scenarios is distinguished by the particular value of the distance between the lugs characterized by parameter lug_distance.

To complete the definition of the **Design Scenario**, we need to inform COSMOSWorks what needs to be reported in these eight steps. Select the **Result Locations** tab in the **Design Scenario** window (Figure 15-6) to access the screen shown in Figure 15-7.

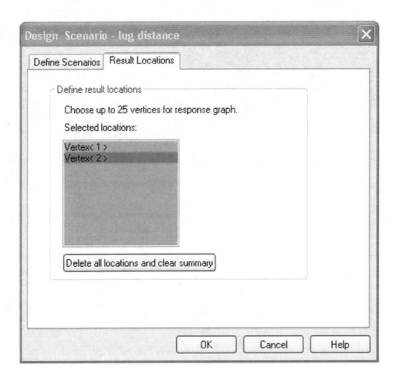

Figure 15-7: Result Locations tab in the Design Scenario window

*As the parameter is modified by the Design Scenario, results for selected locations are recorded. The location must be a vertex. If desired, defined locations can be renamed by right-clicking the item and selecting **Rename** from the associated pop-up menu. Here Vertex <1> and Vertex <2> correspond to points 1 and 2 in Figure 15-1.*

Since only vertexes are allowed as locations in the **Design Scenario** definition, it is now clear why split lines were added to the model geometry.

To run the design scenario, right-click the study folder, open the associated pop-up menu, and select **Run Design Scenario**. Meshing is not required prior to executing the **Design Scenario**. The graphs summarizing the results of all configurations are defined by right-clicking the *Results* folder to open the associated pop-up menu, from which select **Define Design Scenario Graph**.

Figure 15-8 presents the **Graph** window and the corresponding graph showing displacements in locations 1 and 2. Figure 15-9 shows the **Graph** window and the corresponding graph presenting maximum global displacement as a function of distance between lugs.

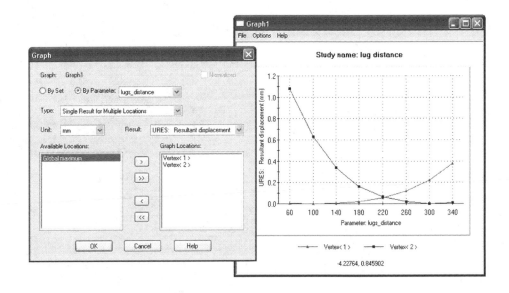

Figure 15-8: Graph definition window and corresponding graph showing displacement in the locations 1 and 2 shown in Figure 15-1.

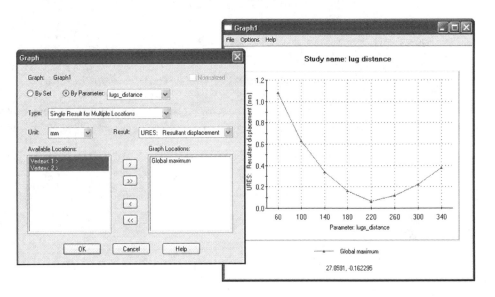

Figure 15-9: Graph definition window and corresponding graph showing global maximum displacement in the model.

Note that the minimum displacement occurs when the lug distance is close to 220mm. In order to find a more precise lug distance corresponding to the minimum deflection value, the parameter increment would have to be refined.

You are encouraged to investigate the other options in the 2D Chart Control Properties window. The options offer ample opportunities to manipulate graph display (Figure 15-10).

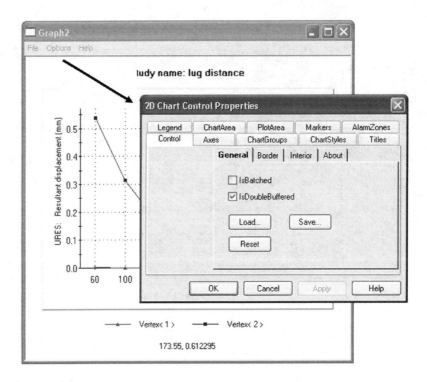

Figure 15-10: 2D Chart Control Properties window

The graph options allow customized graph display.

You may wish to use the same model to run Design Scenario in frequency analysis study to find the distance between lugs that maximizes the first natural frequency.

16: Drop test of a coffee mug

Topics covered

- Drop test analysis
- Stress wave propagation
- Direct time integration solution

Project description

A porcelain mug is dropped from the height of 100mm and lands flat on a horizontal, flat rigid floor (Figure 16-1). We will simulate the impact using COSMOSWorks **Drop Test** analysis.

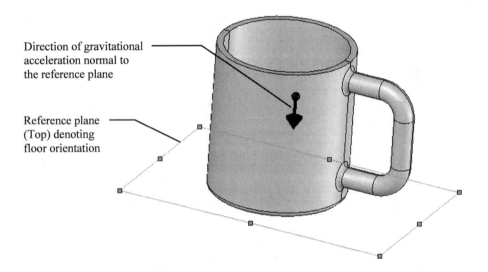

Figure 16-1: Mug landing square on a rigid floor

Procedure

Open the part file called MUG. It has ceramic porcelain material properties already assigned. Note the split lines added to mug geometry.

Create study *drop* specifying **Drop Test** as analysis type and **Solid mesh** as mesh type (only solid mesh is available in **Drop Test** analysis). COSMOSWorks creates two folders in **Drop Test** study: *Setup* and *Result Options* (Figure 16-2).

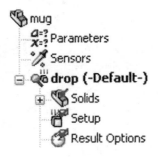

Figure 16-2: *Setup* and *Result Options* folders in Drop Test study

Right-click the *Setup* folder and select **Define/Edit** to open the **Drop Test Setup** window shown in Figure 16-3.

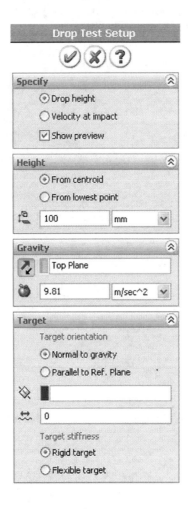

Figure 16-3: **Drop Test Setup** window

Select **Drop height** and **From centroid** and enter 100mm as the drop height. From the fly-out SolidWorks menu (not shown in figure 16-3), select **Top** plane to define the line of action of the gravitational acceleration. Define the direction as shown in Figure 16-3. Enter the magnitude of gravitational acceleration as 9.81m/s^2. Finally, select **Normal to gravity** as **Target Orientation**. After completing the exercise, try experimenting with different **Target Orientation** using **Parallel to Ref. Plane** option in **Drop Test Setup** window.

To ensure that three elements cover the 90° rounds we need to apply mesh control (Figure 16-4).

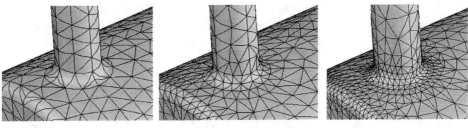

TURN ANGLE 90°   TURN ANGLE 45°   TURN ANGLE 30°

Figure 16-4: Element turn angle

90 turn angle should be avoided in locations where detailed stress results are required.

Apply mesh controls as shown in Figure 16-5 and mesh without **Automatic Transition.**

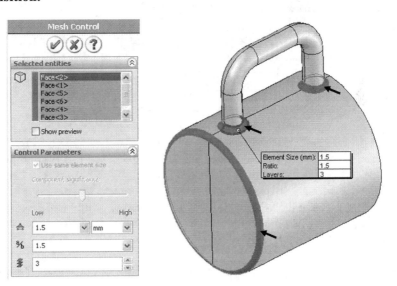

Figure 16-5: **Mesh Control** window

Mesh control of 1.5mm size assures 30° element turn angle in locations indicated by arrows.

Having defined all the required entries in the **Drop Test Setup** window, we now define the results options. Right-click the *Result Options* folder and select **Define/Edit** to open the **Result Options** window (Figure 16-6).

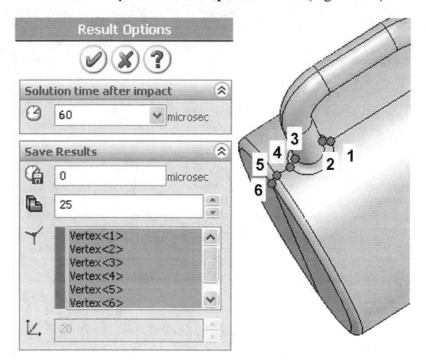

Figure 16-6: **Result Options** window

Locations of vertices used in Results Options. Vertices are created by split lines in the SolidWorks model.

To monitor what happens to the mug during the first 60 microseconds after first impact, enter 60 as **Solution time after impact**. See COSMOSWorks help for more information.

In the **Save Results** area of the **Result Options** window, accept the default 0 (microseconds) meaning that results will be saved immediately after the first impact. Also, accept the default 25 for the **No. of plots**. The solution time is divided into twenty five intervals and full results (available as plots) are saved only for those intervals.

In the vertices field of the **Result Options** window, select the six vertices shown in Figure 16-6. Results for these six vertices will be available in the time history plots.

Note that full results are saved for 25 plots spaced out evenly over a 60 microseconds time period. Results pertaining to time points "in-between" plots are saved only for selected points (here six vertices).

In the last entry in the **Result Options** window we define how many of these partial results are saved. This is done in the **No. of graph steps per plot** field. If we accept the default 20 results, then the total number of data points for each graph (displacement, stress, etc.) is equal to the number of plots times the number of graph steps per plot.

Note that the number of graph steps per plot is not equal to the number of actual time steps. Time steps are selected internally by the solver and may vary as required for the stability of the numerical solution. Accepting the default 20 as the **No. of graph steps per plot** completes **Result Options** window definition.

Because the impact time is very short, it is measured in microseconds. The maximum displacements or stress may occur during the first impact or later when the model is rebounding. A sufficiently long solution time needs to be specified to capture maximum displacement and stress.

Upon completion of the solution, COSMOSWorks creates the following result folders: *Stress*, *Displacement* and *Strain*. To examine a time history graph, right-click the *Results* folder and select **Define Time History Plot** (Figure 16-7).

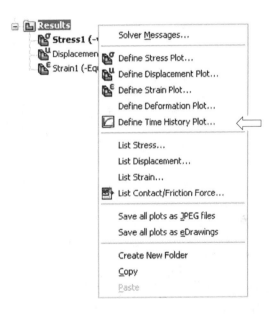

Figure 16-7: Define Time History Plot creates graphs of stress, displacement, velocity and acceleration as function of time.

Plots are available only for previously selected vertices.

To assess if the mug is likely to break, we analyze maximum principal stress P1. Note that maximum principal stress relates to failure of porcelain, which is a brittle material. Select Vertex 2 from **Predefined Locations** and then select the P1 stress to be plotted in MPa (Figure 16-8). For comparison, also create a plot specifying all locations specified in Figure 16-6. Both graphs are shown in Figure 16-9.

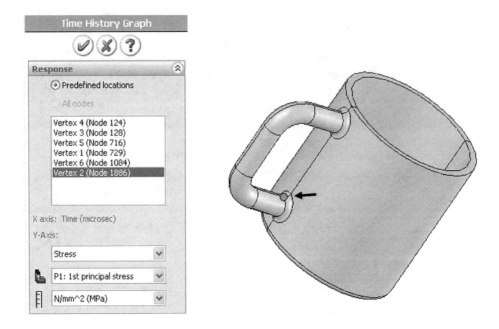

Figure 16-8: Time History Graph definition for vertex 2 location.

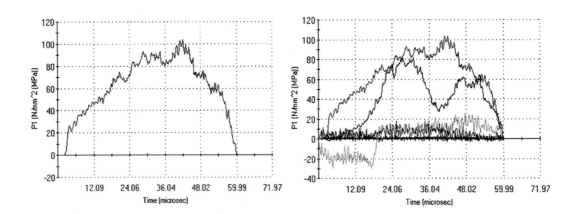

Figure 16-9: Time History of P1 stress in vertex 3 (left) and in all selected locations (right).

Based on the review of **Time History Graph** we find that the highest P1 stress occurs at time of 40 microseconds (move the cursor over the plot to find this time value).

Create a P1 stress plot approximately corresponding to this time point. Right-click the *Stress* folder to open the **Stress Plot** window. In the **Stress Plot** window select the time step that is closest to 40 microseconds (this is time step 17). Stress results are shown in Figure 16-10.

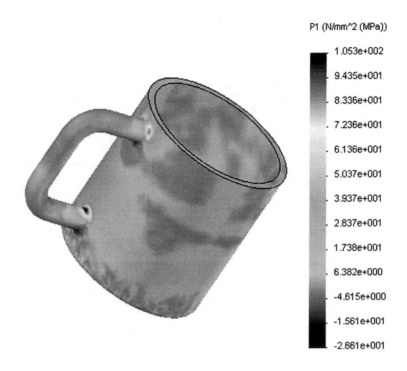

Figure 16-11: Maximum principal stress P1 for time step 17 corresponding to 40.8 microseconds.

Animate this plot to see the dynamic impact.

Will the mug break? **Drop Test** analysis does not directly provide pass/fail results. It is best used to compare the severity of impact for different drop scenarios.

The maximum principal stress is 105 MPa as compared to the ultimate strength of ceramic porcelain which is 172 MPa. This comparison indicates that damage to the mug case is unlikely as long as that mug really lands perfectly flat on the floor.

To see the mug bouncing off the floor as well as stress wave propagating in the model, animate the stress plot. Repeat the study using a longer solution time. If a long enough solution time is used for the analysis, you will see the mug bouncing off the floor more than once and hitting it in different locations.

The **Drop Test** is an analysis intended to model the dynamic impact force of very short durations, since this is when damage is most likely to occur. **Drop Test** analysis takes into consideration inertial effects but no damping. **Drop Test** analysis uses a numerically intensive but stable direct time integration method.

Notes:

17: Selected large displacement problems

Topics covered

- Large displacement analysis
- Creating shell element mesh on the face of a solid

In all previous exercises we assumed that the model stiffness did not change significantly when the model deformed due to the applied load. Consequently, the stiffness only needed to be calculated once, before any load had been applied. This stiffness adequately described model behavior during the entire loading process. The model stiffness did not have to be updated and load could be applied in one single step. The only time we departed from these assumptions was in the analysis of the contact problem.

We will now discuss a few problems where the stiffness does change significantly due to the model displacing under load, and this change is global (not local, as in the contact problem). These problems require a **Large displacements** formulation which is an option in Static study (Figure 17-1).

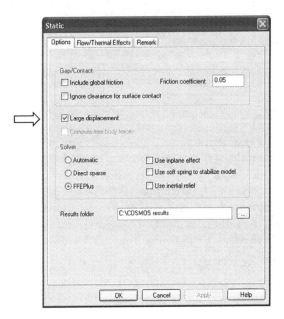

Figure 17-1: Large displacement option in Static study

You may always select the Large displacement option, whether necessary or not. However, this significantly increases solution time.

Open model NL002 and create two static studies with the mesh option **Shell mesh using surfaces** (Figure 17-2). In one of these studies check **Large displacement**.

Using mesh option **Shell mesh using surfaces** we will create a shell element mesh on one side of the solid model. The shell mesh will adequately capture model behavior, avoiding a very large number of solid elements, which would have been required to mesh this thin beam.

Figure 17-2: Static analysis with mesh type Shell mesh using surfaces

This mesh option creates shell elements on selected faces of solid geometry or on surfaces, if they exist in model geometry (not our case).

Since study specifies mesh type as **Shell mesh using surface**, the familiar **Solids** folder is replaced by a **Shells** folder. We need to select which faces will be meshed with shell elements, and then define the element thickness. This process is shown in Figure 17-3.

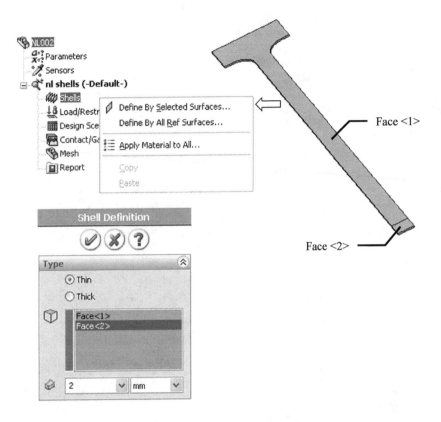

Figure 17-3: Shell definition on the faces of one side of the solid model

Assign 2mm for shell element thickness. This corresponds to the thickness of the solid geometry.

Having preformed the steps shown in Figure 17-3, we see that a **Shell-1** subfolder in the **Shells** folder has been created (Figure 17-4).

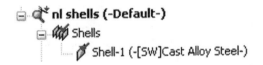

Figure 17-4: Shell-1 subfolder in Shells folder

Material properties are transferred automatically from the SolidWorks model.

Apply a 40N **Normal load** to the small face at the end of the beam and a **Fixed** restraint to the edge at the wide end. Make sure that load and restraint is applied on the side designated for meshing with shell elements. Otherwise, load and restraint won't be transferred from the CAD geometry to the FEA model (Figure 17-5)

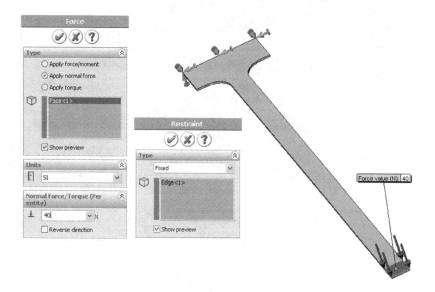

Figure 17-5: Restraint and load applied to the model

Note that Restraint must be applied to the edge of the end face on the side of the shell element and not to the end face (because the end face does not become part of shell element mesh). Load is applied to a small face specifically created in SolidWorks by Split line for a convenient load application.

Now mesh the model with shells 10mm in size and **Automatic Transition** deselected from the **Mesh** options (Figure 17-6).

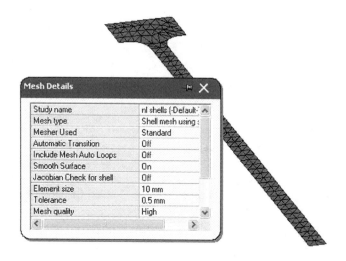

Figure 17-6: Shell mesh created on the side of solid model

Solve the study without the **Large Displacement** option selected and display the displacements plot superimposed on the undeformed model. The solution without the **Large displacement** option displays a warning message (Figure 17-7) that must be acknowledged to complete the solution process.

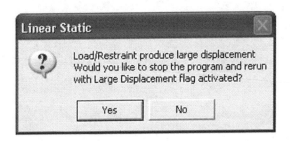

Figure 17-7: A warning message displayed by solver if large displacements are detected and Large displacement option is NOT checked

To obtain linear solution select No.

Now solve the study with **Large displacement** option checked. Solution with **Large displacement** option is a nonlinear solution, which progresses in iterations. Load is applied in automatically determined increments while model stiffness changes due to progressing deformation and is updated during the solution process (Figure 17-8).

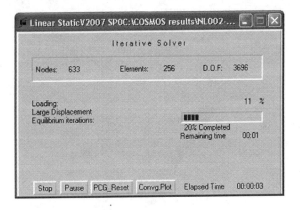

Figure 17-8: Nonlinear solution (with Large displacement option) progresses in iterations while load gradually increases and the stiffness is updated

Because of the iterative nature of the Large displacement solution, the solution time is much longer in comparison to the corresponding linear solution.

Create displacement plots obtained from the nonlinear and linear solutions (Figure 17-9). Use a 1:1 scale of deformation for both plots.

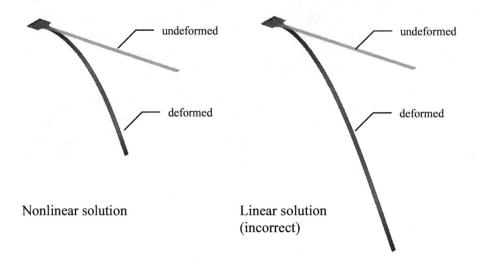

Figure 17-9: Nonlinear solution (obtained with Large displacement option) on the left correctly shows the deformed model. Compare this with the incorrect linear solution on the right

Note that linear solution produces an incorrect pattern of deformation where the tip of the beam travels along a straight line and beam appears to be stretching out.

Now create a von Mises stress plot based on the nonlinear solution using a 1:1 scale of deformation (Figure 17-10).

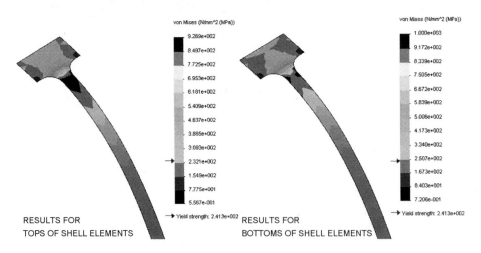

Figure 17-10: Von Mises stress results of the nonlinear solution

Note that the maximum stress on top of the shells (here the compressive side) is 3.9 times higher, and on the bottom of the shells (here the tensile side) it is 4.2 times higher than the yield strength of the model material (241MPa).

In reality, the beam would have yielded long before acquiring the shape shown in Figure 17-9. The "jaggy" shape of the fringes seen in Figure 17-10 and revealed by using discrete rather than continuous fringes, indicates that a more refined mesh would be required for accurate stress results.

Why did the model not yield despite the stresses exceeding the yield strength by 3.8 times? This is because we used the linear material model. The only source of nonlinear behavior that can be modeled with COSMOSWorks 2007 Professional is nonlinearity due to **Large displacement** and even that is done with important simplifications; the load is "ramped up" linearly (there is no other option) and local strain must remain small even though they may add up to globally large model deformations. To account for other sources of nonlinear behavior (e.g. material yielding) and to be able to apply load with a time history, we would need to use COSMOSWorks 2007 Advanced Professional.

Repeat this exercise with a lower load magnitude to find the maximum force magnitude that does not cause yielding.

In the second example we study a large displacement combined with a contact problem. To illustrate this nonlinear contact problem, we analyze the assembly CLIP (Figure 17-11). Even though the clip is just one part, it must be split into two parts to make it into an assembly, because the contacting faces must belong to different parts.

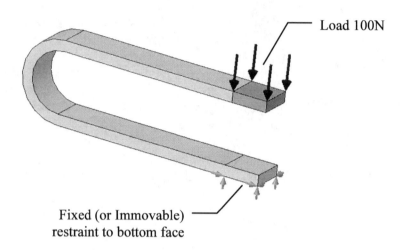

Figure 17-11: The Clip is modeled as an assembly because contacting faces must belong to different parts

Two surfaces, defined in contact pair, will experience large displacement (even though not quite as large as in NL002 example) before contacting each other. Therefore, the **Large displacement** option must be selected in the properties of the **Static** study.

Global **Contact/Gaps** conditions are left at the default setting: **Touching faces: Bonded**. Local **Contact/gaps** conditions between two surfaces likely to come in contact are defined as: **No penetration** (Figure 17-12). Note that we say "likely" because this depends on the load magnitude.

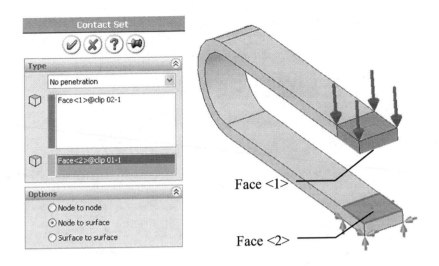

Figure 17-12: Definition of Contact Set

Split lines in the SolidWorks model define small faces that may come in contact. The smaller size of the contacting faces speeds up the solution time.

Mesh the assembly with the default element size, then obtain two solutions: one with **Large displacement** option selected, and the other without. Compare the displacement results obtained from these two studies (Figure 17-13).

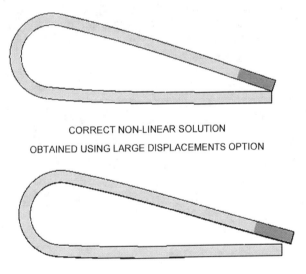

Figure 17-13: Correct displacement results produced with the **Large displacement** option selected (top) and incorrect displacement results produced without **Large displacement** option (bottom)

The Gap did not close properly in the solution executed without the Large displacement option.

Note that the element size is much too large in the contact area to produce meaningful contact stress results (Figure 17-14).

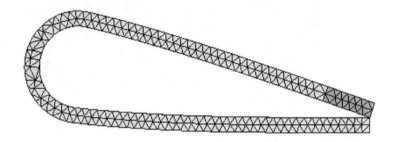

Figure 17-14: The large element size in the contact area prevents meaningful analysis of contact stresses

The two pervious exercises (NL002 and CLIP) required **Large displacement** option because displacements were indeed large. However, there are problems where displacements are small, yet they significantly change model stiffness. These problems still require solutions with the **Large displacement** option selected. In fact, the term **Large displacement** may be confusing because it implies that this option is only applicable if displacements are large.

We will demonstrate that displacements don't have to be large in order to require a solution with the **Large displacement** option selected. Open model ROUND PLATE, which is a thin plate subjected to pressure. Un-suppress the cut (the last feature in the SolidWorks Feature Manager) to work with a slice, as shown in Figure 17-15.

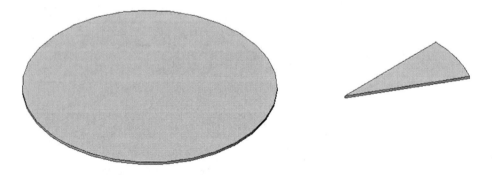

Figure 17-15: The full ROUND PLATE model (left) and a section obtained by un-suppressing the cut.

Because of axial symmetry of the geometry, loads and restraints, an arbitrary slice with applied boundary conditions correctly represents the plate's response to pressure.

Create a static study with a **Solid mesh** and apply **Symmetry** restraints to both radial faces created by the cut. Next, apply a **Fixed** restraint to the cylindrical face on plate circumference and a pressure of 300,000Pa to the top face (Figure 17-16).

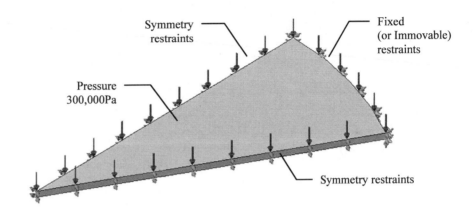

Figure 17-16: Load and restraint applied to the ROUND PLATE model

Obtain two solutions: with and without **Large displacement** option and compare the resultant displacement results (Figure 17-17).

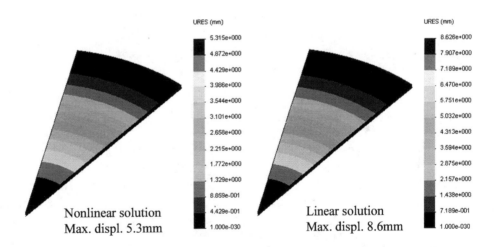

Figure 17-17: Displacement results obtained with the Large displacement option (nonlinear solution) (left) and displacement results obtained without the Large displacement option (linear solution) (right).

A nonlinear model deforms less than a linear model, meaning the nonlinear model is stiffer than the linear model.

A flat plate under pressure is a classic case where the assumption of small displacements leads to erroneous results. The analysis requires a **Large displacement** option even though displacements are small in comparison to the size of the model.

The need to use the **Large displacement** option is due to the change of shape (from flat to curved) altering the mechanism of resisting the load; a deformed plate is able to resist pressure with membrane (tensile) stress additionally to the original bending stress (Figure 17-18).

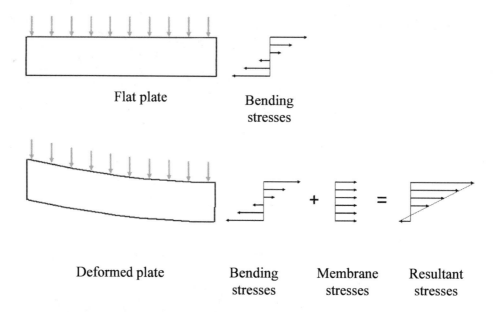

Figure 17-18: Pressure resisting mechanism in undeformed plate (top) and in deformed plate (bottom)

To account for the change of plate stiffness that takes place due to deformation (even though that deformation is small), the stiffness must be updated during the deformation process. This is possible only if a nonlinear analysis is executed. The linear analysis only takes into account the initial bending stiffness, which is the reason why the linear model is softer that the corresponding nonlinear model.

We will illustrate the same problem with one more example, similar to LINK described in Chapter 5. Open model LINK02 and note that the material properties of ABS have been assigned to this model. Due to material low stiffness, the model will experience large deformation which will make it easy to show the difference between the correct nonlinear (**Large displacement**) solution and incorrect linear solution. Apply 150N to each one of two top faces, and hinge restraints to the "eyes" (Figure 17-19).

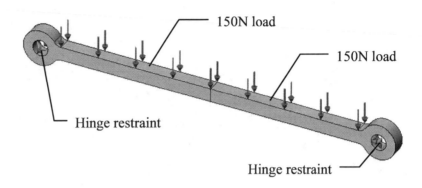

Figure 17-19: Loads and restraints on LINK02 mode

Split lines are here for convenient probing of results.

Obtain nonlinear and linear solutions and compare von Mises stress results from both solutions (as shown in Figure 17-20).

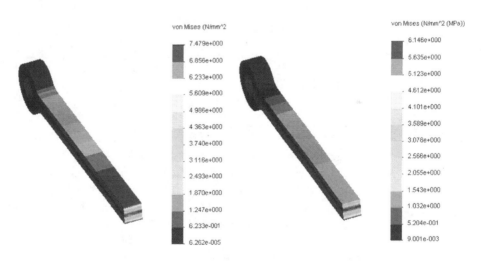

LINEAR SOLUTION MODELS ONLY BENDING STRESSES; THIS IS INCORRECT

NON-LINEAR SOLUTION MODELS CORRECTLY MODELS BENDING AND MEMBRANE STRESSES

Figure 17-20: Von Mises stress results for nonlinear (top) and linear (bottom) solutions for the same problem.

The absence of membrane (tensile) stresses in the linear solution is illustrated by symmetric distribution of von Mises stresses in the section stress plot (left).

The von Mises section plot obtained in nonlinear (large displacement) solution shows non-symmetric distribution stresses proving the presence of membrane (tensile) stresses (right). Refer again to Figure 17-18 for explanations on how bending and membrane stress add.

The difference between linear and nonlinear solutions is schematically shown in Figure 17-21.

Figure 17-21: The nonlinear solution correctly models displacements and stresses in hinge supported link where the distance between hinges can not change (top). The linear solution can only model the configurations where one of hinges is floating (bottom).

The nonlinear solution correctly models displacements and stresses in a hinge supported link as shown at the top of Figure 17-21. The linear solution, not being capable of modeling membrane stress that develops during the deformation, models the link as if one of hinges had a floating support (symbolically shown in Figure 17-21 as rollers). However, horizontal displacements are not modeled in linear solutions, even if floating support is modeled. Therefore, a linear model cannot distinguish between the two configurations shown in Figure 17-21.

18: Mixed meshing problem

Topics covered

- Using solid and shell elements in the same mesh
- Mixed mesh compatibility

While in previous exercises we have used different types of meshes such as solid and shell meshes, we have never used them in the same model. This exercise introduces the use of different mesh types within the same model.

Open the assembly model WHEEL. Note that the 'bulky' hub and rim are connected by thin spokes (Figure 18-1). Since the spokes are thin, using a solid element mesh to model them would require a large number of small solid elements. Therefore, to reduce the problem size, hub and rim will be meshed with solid elements and spokes with shell elements.

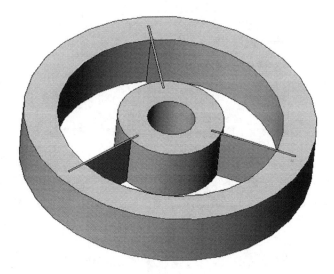

Figure 18-1: A "bulky" rim and hub, suitable for solid element meshing, are connected by three thin spokes, which are best meshed with shell elements.

Shell elements can be created by meshing surfaces (in our model there are no surfaces) or faces of solid geometry. In this exercise we will create shells on the faces of solids representing spokes.

Create a **Frequency** study with **Mixed mesh** and notice that the study folder contains two folders: **Solids** and **Shells**. We now need to select which faces in the assembly will be meshed with shell elements. Right-click the **Shells** folder and select **Define By Selected Surfaces** to open the **Shell definition** window. Use **Thin** shell formulation, enter 1.5mm as shell thickness (this is the thickness of the spoke) and select one side of each spoke to designate it for meshing with shell elements. This process is explained in Figure 18-2.

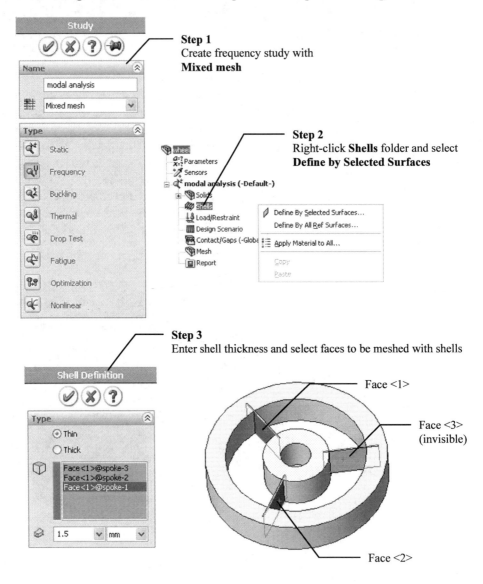

Figure 18-2: Three steps in designating faces for meshing with shell elements. Shell thickness is 1.5 mm.

Note that shells will be created on one side of each spoke. We consider this as an acceptable modeling simplification.

A very important part of this exercise is definition of contact conditions. In a **Mixed mesh** study, all global and component contact conditions are ignored and we must assure connectivity between assembly components by defining a local contact condition. Switch to the exploded view and define three **Contact sets** as explained in Figure 18-3.

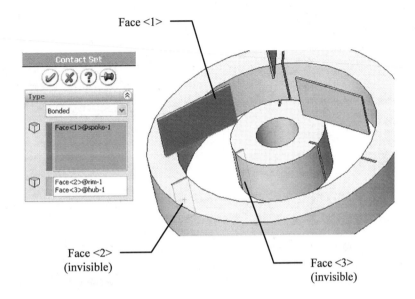

Figure 18-3: The Bonded contact condition is defined between the face of the spoke, which will be meshed with shell elements (Face<1>), and the faces of the rim (Face<2>) and the hub Face<3> that touch the spoke.

In total, three contact conditions need to be created in the model.

By defining local contact conditions, we assure connectivity of all components. The model is now ready for meshing. Mesh it with the default element size. The mixed mesh is shown in Figure 18-4.

Before solving, apply a **Fixed** restraint to the hole in the hub.

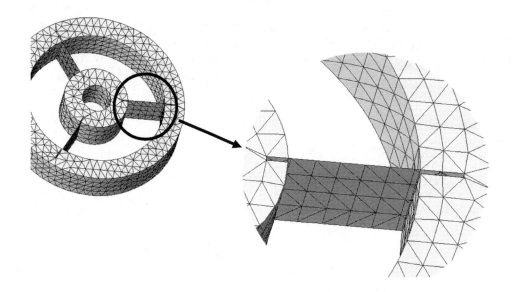

Figure 18-4: Mixed mesh: rim and hub meshed with solids while spokes are meshed with shells.

Note that shell mesh attaches to the side of slots in the rim and in hub. This is the result of modeling simplifications that we deem acceptable for frequency analysis. This would not be acceptable for stress analysis.

Deformation pattern of the first mode of vibration is shown in Figure 18-5.

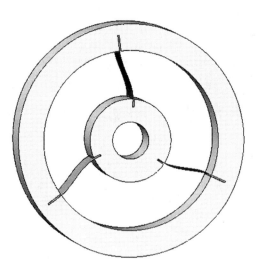

Figure 18-5: Deformation pattern corresponding to the first mode of vibration.

19: Static analysis of a weldment using beam elements

Topics covered

- Different levels of idealization implemented in finite elements
- Preparation of SolidWorks model for analysis with beam elements
- Beam elements and truss elements
- Analysis of results using beam elements
- Limitations of analysis with beam elements

Project description

A Roll-Over Protective Structure (ROPS) is used to protect an operator of heavy equipment in case of roll-over. A simple ROPS consists of eight hollow square tubes 3 x 3 x 0.25 (Figure 19-1). It is created in SolidWorks as a weldment with three structural members comprised of eight beams.

We wish to find displacements and stresses of this structure under the load of 5,000lbs with all four legs restrained.

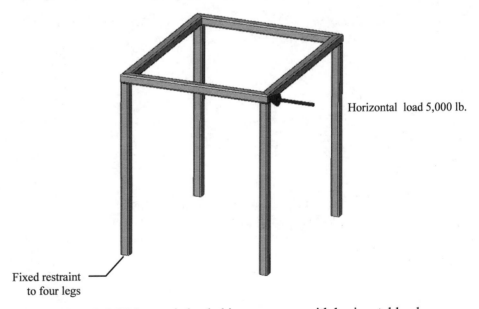

Figure 19-1: A ROPS cage is loaded in one corner with horizontal load 5,000lbs; all legs are restrained.

The tube cross section and details of corner treatment and trims are shown in Figure 19-2.

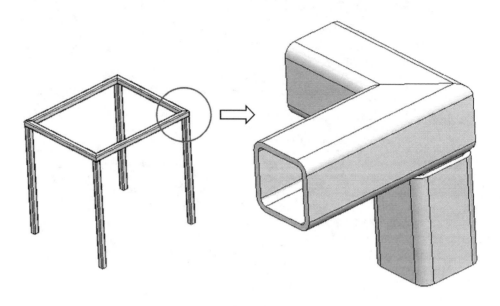

Figure 19-2: A detail of the corner; all tubes are 3 x 3 x 0.25" with 0.5" radius

Corner treatments and trims are applied in the SolidWorks model using Weldment tools; weld bead is not modeled.

Due to thin walls and complicated geometry in corners, this model is not suitable for meshing with solid or shell elements. Even if we were ready to accept long meshing and solution times, stress results in the corners would be useless because of numerous sharp re-entrant edges causing numerous stress singularities.

To avoid these problems, the model can easily be meshed and analyzed with beam elements. Before we proceed with analysis we need to explain what beam elements are and how they compare with solid and shell elements.

The difference between solid, shell and bam elements and summarized in the following table.

Element type	Idealizations made to geometry intended to be meshed with this element	Assumptions on stress distribution in the element
Solid	None; solid elements are created by meshing 3D solid geometry.	No assumptions on stress distribution need to be made
Shell	Shell elements are created by meshing surface. This can be a "stand alone" surface or a face of a solid geometry. Thickness is not present in the geometry (this dimension is dropped) and must be entered as a numerical value in the shell element definition.	Assumptions on stress distribution across thickness are made. In-plane stresses are assumed to be distributed linearly across the thickness. Shear stresses are either assumed to have uniform distribution across element thickness (thin shell formulation) or to have parabolic distribution (thick shell formulation).
Beam	Curves are used to create beam elements. Curves represent geometry with two dimensions dropped. In CAD terminology this is wire frame geometry.	Assumptions on stress distribution must be made in two directions perpendicular to the curve. These assumptions are the same as in the beam theory: bending stresses are distributed linearly in both directions, and axial and shear stresses are constant.

In summary, solid elements are a natural choice for meshing models with approximately the same size in all three dimensions, while shells are the natural choice to mesh sheet metal models, and beams are the natural choice to mesh structural members.

Beam cross section geometry is only used to define beam element properties such as the area and second moments of inertia of beam cross-section. Beam cross section does not become part of finite element model.

The information about beam cross sections is retrieved from the SolidWorks model which works only if SolidWorks model is created as a Weldment. It is important to understand that solid geometry of a Weldment is not meshed when beam elements are used. What is meshed is the underlying wire frame geometry (Figure 19-3).

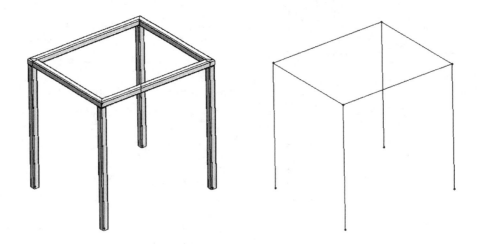

Figure 19-3: Solid model and the underlying wire frame geometry

Solid geometry is used only to define beam cross sections. Beam elements are created by meshing curves (wire frame). You may think of beam elements as lines with assigned beam cross section properties. Currently, only straight lines may be meshed with beam elements.

Corner treatments and trims have no importance in beam element model (Figure 19-4).

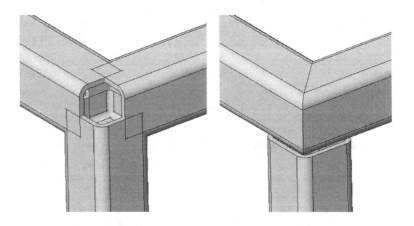

NO CORNER TREATMENT, NO TRIM   CORNER TREATMENT AND TRIM APPLIED

Figure 19-4: Corner treatment and trims have no importance in beam element model. Both geometries will produce the same finite element model when meshed with beam elements.

Procedure

Having examined the ROPS part, move to COSMOSWorks and create a study specifying **Beam mesh** (Figure 19-5).

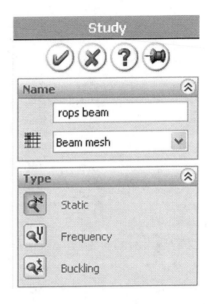

Figure 19-5: Study definition with beam elements

Note that only Static, Frequency and Buckling studies are available for a model using beam elements.

Right-click the automatically created *Beams* folder and select **Treat all structural members as beams** (Figure 19-6).

Figure 19-6: Designation of structural members to be treated as beam elements

This creates eight beams in the *Beam* folder and a **Joint group**, which is used to define connectivity between beams (Figure 19-7).

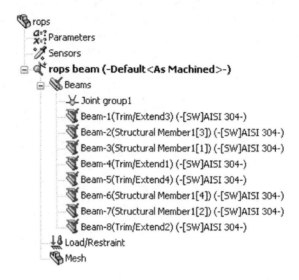

Figure 19-7: Eight beams are created in Beams folder. Also created is Joint group

Material can be imported from SolidWorks model or defined in COSMOSWorks individually to each beam or to all beams.

Right-click **Joint group1** in the **Beams** folder and select **Edit** from a little pop-up menu to open the **Edit joints** window. Accept the default selection **All beams** and click **Calculate** to create joints automatically (Figure 19-8).

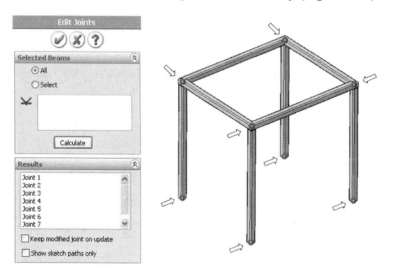

Figure 19-8: Joints are automatically created between beams and at the free ends of beams. Arrows indicate joint positions.

The concept of joints between beams is better conveyed if **Show sketch paths only** is selected in the **Edit joints** window (Figure 19-9).

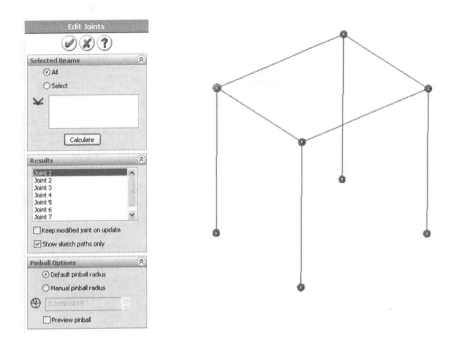

Figure 19-9: Joints are created between line sections of the wire frame geometry

Position of joints is indicated by "pinballs".

Joints (or beam ends) are connected to each other only if they are contained within the pinball radius, which can be changed in **Edit joints** window. It is recommended that beam ends intended to be connected are made coincident. Non-coincident beam ends can still be connected if they fall inside the pinball, but this results in a "patch-up" beam element mesh (Figure 19-10).

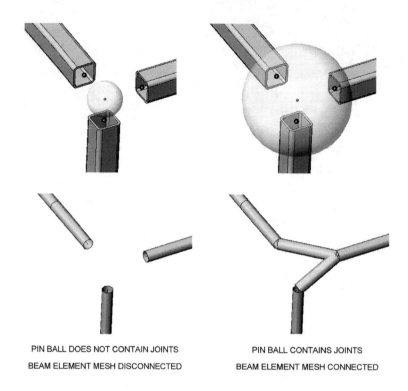

Figure 19-10: Joints (beam ends) are connected only if they are contained within the Pinball. Beam elements are shown as thin tubes.

Disjoined structural members can be connected if their ends fall within the volume of the pinball (top right). However, the beam element mesh (bottom right) will then contain automatically created connecting elements. This may create unpredictable results. Beam elements are graphically depicted as round tubes, even though in fact beam elements are just lines. The tube diameter is always the same, regardless of the actual cross-section size, shape and orientation.

Apply restraints to all four joints at the free ends (at the bottom) of vertical members (Figure 19-11)

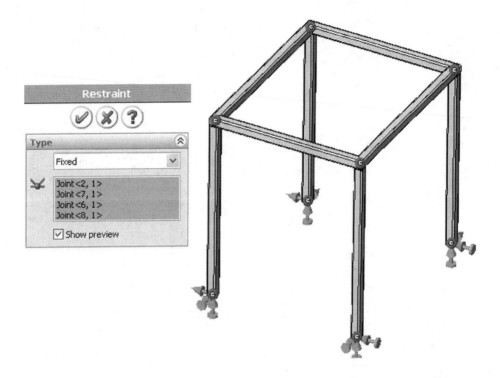

Figure 19-11: Fixed restraint applied to the free ends of vertical members

Note that beam elements have six degrees of freedom per node and, therefore, can distinguish between Fixed and Immovable restraint. Here, we need to apply a Fixed restraint.

Apply 5,000 lb load to the corner joint, as shown in Figure 19-12. Remember that restraints and loads can be only applied to joints.

Figure 19-12: Force load applied to the corner joint

Note that beam elements have six degrees of freedom per node and, therefore, they can be loaded with force as well as with moment load.

Now create the beam element mesh, noting that there are no user controlled mesh parameters. The beam element mesh is shown in Figure 19-13.

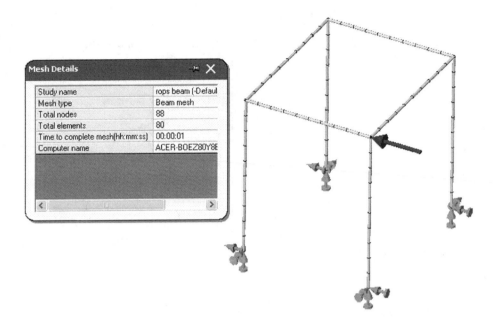

Figure 19-13: Beam element mesh is created from curves (here straight lines) used in the SolidWorks model to define Structural Members.

There are 80 elements and 88 nodes. Count them, noting that in each corner there are three coincident nodes.

Obtain solution, right-click on *Results* folder and select **Solver Messages** to read the number of nodes, elements, degrees of freedom and the solution time (Figure 19-14).

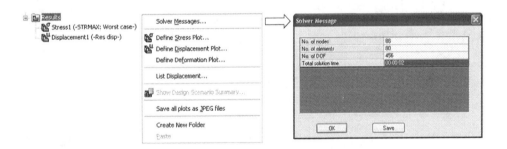

Figure 19-14: Number of nodes is 88, number of elements is 80 and the number of degrees of freedom is 456.

Verify these numbers by counting elements and nodes. Remember that restrained nodes do not contribute to the total number of degrees of freedom and that three coincidental nodes in all corners share the same 6 degrees of freedom. Each straight line segment of wire frame is meshed with 10 beam elements.

Create a displacement results plot with the undeformed model superimposed on the displacement plot (Figure 19-15) and stress plot (Figure 19-16).

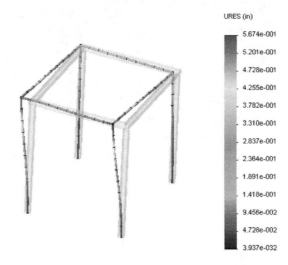

Figure 19-15: Resultant displacement results of ROPS model

Note that while the undeformed model is shown as solid CAD geometry, the displaced model is shown as an assembly of "tubes".

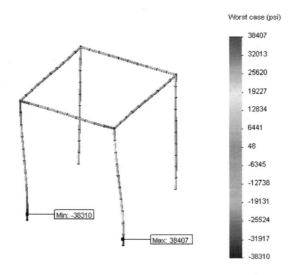

Figure 19-16: Stress plot of "worst case" stresses showing the location of maximum (tensile) stress and minimum (compressive) stress. The default stress plot shows the "worst case" stress.

The maximum tensile stress is almost 38.5ksi; the maximum compressive stress is 38.5 ksi (shown as -38.5 ksi).

The stress results, compared with material yield strength of 30ksi, indicate that yielding will take place in the ROPS weldment. The results do not provide a direct answer whether or not the structure will collapse.

Note that the "worst case" stress is NOT von Mises stress. To understand what "worst case" stress is, we need to review all stress results options available for beam elements.

The software provides the following options for viewing stresses (refer to Figure 19-17):

- Axial: Uniform axial stress = P/A

- Bending in local direction 1: Bending stresses due to moment M1 about axis 1.

- Bending in local direction 2: Bending stress due to moment M2 about axis 2

- Worst case

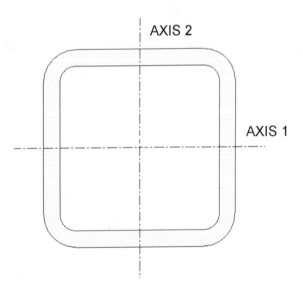

Figure 19-17: Positions of axis1 and axis2 for beam cross section used in this exercise.

Axis1 and Axis2 cross the centroid of the cross section. The second moment of inertia of this cross section about each axis is the same due to double symmetry of this cross section.

Worst case stress is calculated by combining axial stress and bending stresses due to moments M1 and M2. This is the recommended stress to view.

In general, the software calculates 4 stress values at the extreme fibers of each end. When viewing worst case stresses, the software shows one value for each beam segment. This value is the largest in magnitude out of the 8 values calculated for the beam segment.

For better understanding of beam elements we will review another example shown in Figure 19-18. Open part TRUSS.

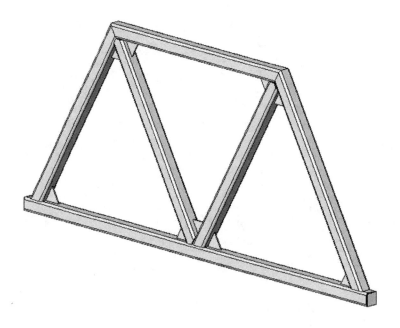

Figure 19-18: Truss is made out of the same rectangular hollow tube as the one used in the previous example.

The SolidWorks model includes corner treatments, trims and details such as gussets and end caps. These details are not translated into the finite element model.

Define a static study with beam element mesh. Apply fixed restraints and load as shown is Figure 19-19. Remember that loads and restraints can be applied only to joints. Mesh with beam elements noting that details such as gussets and end caps do not become part of the mesh. Obtain solution and review results.

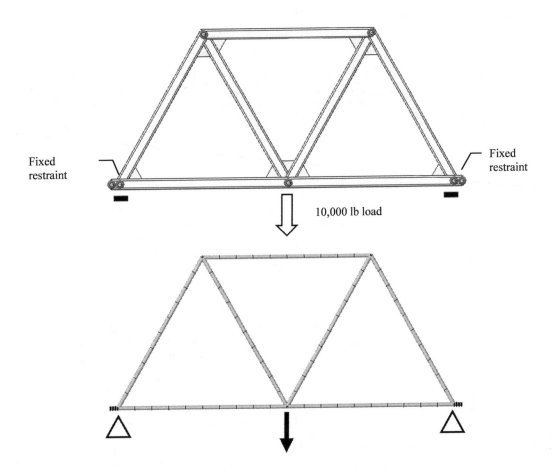

Figure 19-19: Restraints and loads applied to joints (top) and finite element mesh with restraints and loads applied to element nodes (bottom).

Each section of structural member, regardless of its length, is meshed with 10 elements.

Now create an identical study to the one you have just solved. The study must be created "by hand" as copying studies does not work for studies with beam elements. Select all beams in the *Beams* folder and right-click to open a pop up menu. Select **Edit definition** to open the **Apply/Edit beam** window. Select **Truss** in the **Apply/Edit beam** window (Figure 19-20).

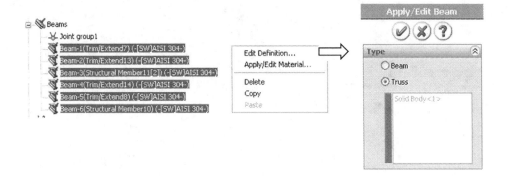

Figure 19-20: All beams are now defined as trusses

This can be done for all beams or individually for selected beams.

This redefines connectivity between beams from rigid to pin joints. While abeam can be loaded with any combination of forces and moments, a truss can be only loaded with axial force. A truss behaves as a tension/compression spring and is meshed with only one element. Verify this by examining meshes in both studies.

Compare displacement and stresses results between two studies (Figure 19-21).

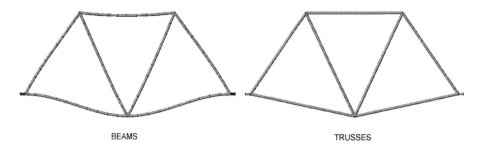

Figure 19-21: The deformation pattern of model with beam elements and truss elements.

Bending of structural members observed in the beam element model proves that beam elements are rigidly connected to each other and transmit bending moments. On contrary, truss elements are connected by pin joint. They can not transmit bending and remain straight.

To complete this exercise, verify that 'worst case" stress and axial stress are the same for the truss element model. This is because axial stress is the only stress component present in truss elements.

20: Miscellaneous topics

Topics covered

- Mesh quality
- Solvers and solvers options
- Displaying mesh in result plots
- Automatic reports
- E drawings
- Non-uniform loads
- Bearing load
- Frequency analysis with pre-stress
- Shrink fit analysis
- Connectors
- Pin connector
- Bolt connector

The analysis capabilities of COSMOSWorks go beyond those we have discussed so far. Readers are now sufficiently familiarized with COSMOSWorks to explore options and topics we have not covered. To aid this effort, in this Chapter we review a variety of topics that have not been addressed in previous exercises.

Selecting the automesher

You can select the **Standard** or **Alternate** automesher in the Preferences window under the **Mesh** tab (Figure 2-13). The Standard automesher is the preferred choice. It uses the Voronoi-Delaunay meshing technique and is faster than the alternate automesher.

The **Alternate** automesher uses the Advancing Front meshing technique and should be used only when the **Standard** automesher fails, even when various element sizes are tried. The **Alternate** mesher ignores mesh control and automatic transition settings.

Mesh quality

The ideal shape of a tetrahedral element is a regular tetrahedron. The aspect ratio of a regular tetrahedron is assumed as 1. Analogously, an equilateral triangle is the ideal shape for a shell element. During meshing, elements are mapped on model geometry. This distorts the element shape. The further the element departs from its original shape, the higher the aspect ratio becomes (Figure 20-1). An aspect ratio that is too high causes element degeneration, which negatively affects the quality of the results.

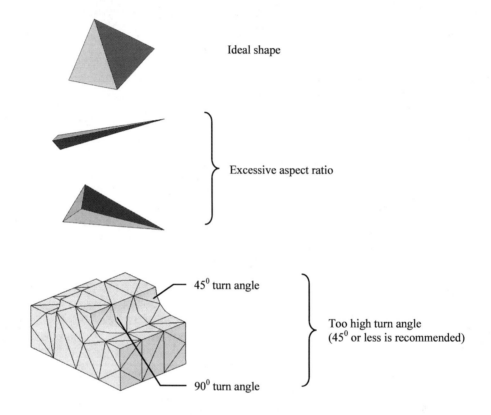

Figure 20-1: Tetrahedral element shapes: ideal and after mapping

A tetrahedral element in the ideal shape (top) has as aspect ration of 1. "Spiky" and "flat" elements shown in this illustration (middle) have excessively high aspect ratios. "Concave" elements (bottom) have excessive turn angles.

The aspect ratio of a perfect tetrahedral element is used as the basis for calculating the aspect ratios of other elements. While the automesher tries to create elements with aspect ratios close to 1, the nature of geometry sometimes makes it impossible to avoid high aspect ratios (Figure 20-2).

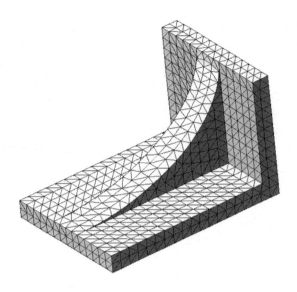

Figure 20-2: Mesh of an elliptical fillet

Meshing an elliptical fillet creates highly distorted elements near the tangent edges.

A failure diagnostic can be used to spot problem areas if meshing fails. To run a failure diagnostic, right-click the *Mesh* icon, which opens the associated pop-up window, and select **Failure Diagnostic ...** (Figure 20-3). From the same menu you may select **Create Mesh plot**, showing the mesh itself, or **Mesh Quality** measures, such as **Aspect Ratio** and **Jacobian**.

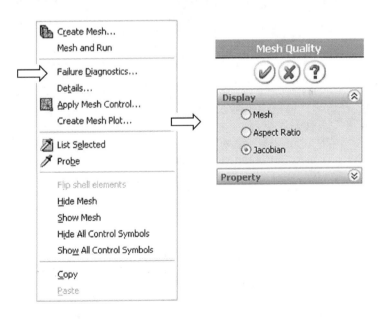

Figure 20-3: Failure Diagnostic and Mesh Quality window can be accessed from this menu

Try creating mesh quality plots for some of the completed exercises specifying Aspect Ratio and Jacobian.

Meshing difficult geometries may sometimes result in degenerated elements without any warning. If mesh degeneration is only local, then we can simply not look at results (especially the stress results) produced by those degenerated elements. If degeneration affects large portions of the mesh, then even global results cannot be trusted.

Solvers and solvers options

In finite element analysis, a problem is represented by a set of algebraic equations that must be solved simultaneously. There are two classes of solution methods: direct and iterative.

Direct methods solve the equations using exact numerical techniques, while iterative methods solve the equations using approximate techniques. With the iterative method, a solution is approximated with each iteration and the associated errors are evaluated. The iterations continue until the errors become acceptable.

Direct Sparse solver (usually slower) uses direct solution technique; FFEPlus uses iterative technique. These solvers are available with three solver options (Figure 20-4).

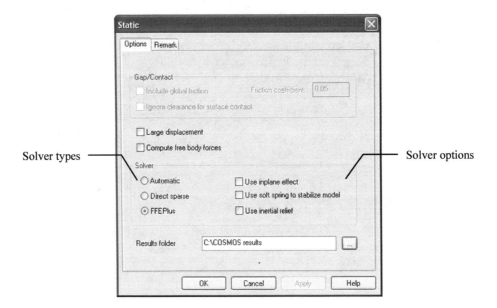

Figure 20-4: Different Solvers and Solver options are available in COSMOSWorks

Both solvers give comparable results if the required options are supported. It is generally recommended to use the **Automatic** option to select the solver automatically based on user specified solver options.

If a solver requires more memory than available, then disk space is used to store and retrieve temporary data. When this situation occurs, a message is displayed that says the solution is going out of core and the solution progress slows down very significantly. Three solver options are available:

Option	Purpose
Use in plane effect	In a static analysis, use this option to account for changes in structural stiffness due to the effect of stress stiffening (when stresses are predominantly tensile) or stress softening (when stresses are predominantly compressive).
	In a frequency analysis, use this option to run a pre-stress frequency analysis.
Use soft springs to stabilize the model	Use this option primarily to locate problems with restraints that result in rigid body motion. If the solver runs without this option selected and reports that the model is insufficiently constrained (an error message appears), the problem can be re-run with this option selected (checked). Insufficient restraints can then be detected by animating the displacement results.
	An alternative to using this option is to run a frequency analysis, identify the modes with zero frequency (these correspond to rigid body modes), and animate them to determine in which direction the model is insufficiently constrained.
Inertial relief	Use this option if a model is loaded with a balanced load, but no restraints. Because of numerical inaccuracies, the balanced load will report a non-zero resultant. This option can then be used to restore model equilibrium.

Several options are available when solving contact problems: **Include global friction, Ignore clearance for surface contact** and **Friction coefficient**. Those options are defined in the study properties, as shown in the table below.

The three options are described as follows:

Option	Purpose
Include global friction	If selected (checked), friction between contacted surfaces is considered.
Ignore clearance for surface contact	Use this option to ignore the initial clearance that may exist between surfaces in contact. The contacting surfaces start interacting immediately without first canceling out the gap
Large displacement	See Chapter 17
Compute free body forces	Enable probing of forces and moments transmitted by nodes.

Displaying mesh in result plots

The default brightness of **Ambient** light, defined in SolidWorks Manager in the **Lighting** folder, is usually too dark to display mesh, especially a high-density mesh. A readable display of mesh (Figure 20-5) requires increasing the brightness of ambient light.

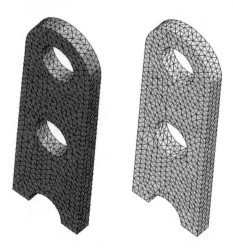

Figure 20-5: Mesh display in default ambient light (left) and adjusted ambient light (right)

A finite element mesh is displayed with ambient light brightness suitable for a CAD model (left), and with the brightness adjusted for the displaying mesh (right).

Automatic reports

COSMOSWorks provides automated report creation. After a solution completes, the *Report* folder contains an automatically created report. To edit it, right-click the report icon to open the **Report** window (Figure 20-6).

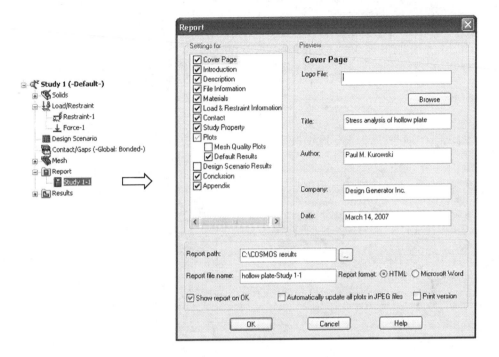

Figure 20-6: *Report* folder

A Report folder may contain several reports. A report can be created only after an analysis has been completed.

The report is created is a few steps and contains all plots from the result folders.

E drawings

Each result plot can be saved in various graphic formats, as well as in SolidWorks eDrawing format. The eDrawing format offers a very convenient way of communicating FEA results to people who do not have COSMOSWorks.

Non-uniform loads

We will illustrate the use of non-uniformly distributed loads with an example of hydrostatic pressure acting on the walls of a 1.95m deep tank, presented in the SolidWorks part file called NON-UNIFORM LOAD. Note that this model uses meters for the unit of length. The pressure magnitude expressed in [N/m^2] follows the equation p = 10,000x, with x being the distance from the top of tank (where the coordinate system *cs1* is located). The pressure definition requires selecting the coordinate system and the face where pressure is to be applied. The formula governing pressure distribution can then be entered in, as shown in Figure 20-7.

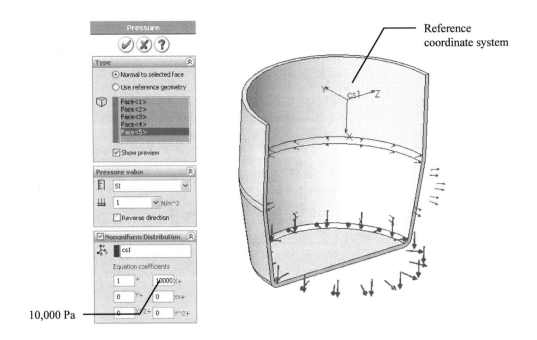

Figure 20-7: A water tank loaded with hydrostatic pressure requires linearly distributed pressure

This illustration uses a section view. Note that the vector lengths correspond to pressure magnitudes that vary with x coordinates of cs1 coordinate system.

You may wish to complete this exercise by applying a **Fixed** restraint to the bottom of the tank and material of your choice. The tank geometry makes it suitable for meshing with shell elements using mid-surfaces.

Deformation plot of the third mode of vibration is shown on the cover of this book.

Bearing load

A bearing load can be used to approximate contact pressure applied to a cylindrical face without modeling a contact problem. As seen in Figure 20-8, the size of contact must be assumed (guessed) as indicated by the split face. This definition requires a coordinate system on which the *z*-axis is aligned with the axis of the cylindrical face. See the part file BEARING LOAD for details.

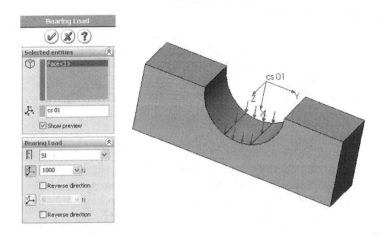

Figure 20-8: 10,000 N resultant force applied as bearing load to a split face

The pressure distribution over the selected face follows the sine function.

Frequency analysis with pre-stress

A frequency analysis of rotating machinery most often must account for stress stiffening. Stress stiffening is the increase in structural stiffness due to tensile loads. We will illustrate this concept with the example of a helicopter blade. Open part ROTOR which comes with assigned material properties and two defined COSMOSWorks studies: *no preload* and *preload*.

Loads definition is not required in **Frequency** analysis. However, if loads are defined, their effect will be considered. Direct sparse is the only solver that accounts for the effect of loads in **Frequency** analysis. If the **Automatic** solver option is selected in properties of the study, the solution switches to **Direct sparse**.

A centrifugal load has been defined as shown in Figure 20-9. An axis or a cylindrical face is required as a reference to define centrifugal load. Review the restraint, which is the same in both studies. Since ROTOR has three identical blades, un-suppress the cut (the last feature in the SolidWorks Feature Manager) to work with geometry containing one blade only. Apply **Symmetry** restraints to faces created by the cut and **Fixed** restraints to the section of central hole.

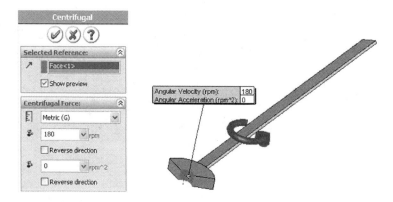

Figure 20-9: **Centrifugal** load window with centrifugal force defined.

A centrifugal load resulting from angular velocity of 180 rpm is applied to the model, simulating the effect of rotation about the axis of the cylindrical face. Angular acceleration can also be defined. The load symbol is located in the center of mass of the model even though rotation takes place about the selected axis.

Solve both studies and compare frequency results (Figure 20-10) without and with pre-stress effect.

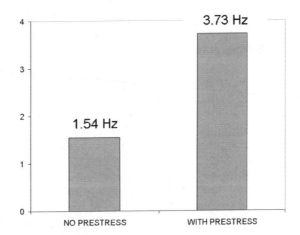

Figure 20-10: Frequency of the first mode of vibration without and with considering the effect of centrifugal load.

As shown in Figure 20-9, the presence of preload significantly increases the frequency of vibration. In the first mode the frequency increases 2.4 times.

The opposite effect (decrease in the natural frequency) would be observed if, hypothetically, the rotor blades were subjected to a compressive load.

For more practice, try conducting a frequency analysis of a beam under a compressive load. You may re-use the ROTOR model. The higher the compressive load, the lower will be the first natural frequency. The magnitude of the compressive load that causes the first natural frequency to drop to 0 (zero) is the buckling load. This is where frequency and buckling analyses meet!

Shrink fit analysis

Shrink fit is another type of **Contact/Gaps** condition. We use it here to analyze stresses developed as a result of an interference (press fit) between two assembly components. Open the SolidWorks model SHRINK FIT. The definition of the shrink fit condition is shown in Figure 20-11. Review this model for definitions of restraints, supports and contact conditions.

Note that the contact condition does not include friction, therefore the inside cylindrical face of the pressed in component has been restrained in circumferential and axial directions to prevent rigid body motions.

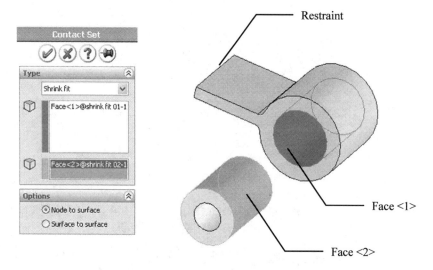

Figure 20-11: Cylindrical face <2> has a larger diameter than cylindrical face <1>. Solving the model with Shrink Fit contact condition eliminates this interference.

Exploded view should be used to select the interfering faces.

Apply a restraint to the "tail" of housing as shown in Figure 20-11 and restraints to the hole of the shaft, shown in Figure 20-12. This is necessary to eliminate rigid body motions of the shaft.

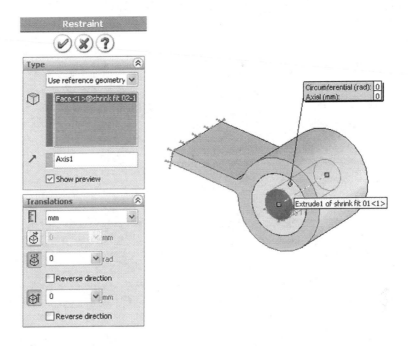

Figure 20-12: Restraints applied to the hole are required to eliminate rigid body motions.

Restraint is applied in circumferential and axial directions. Radial direction is free in order not to interfere with shrinking of the hole due to press-in (shrink fit).

Mesh model with default element size 1.5mm and obtain solution. Display contact stress plot using exploded view (Figure 20-13).

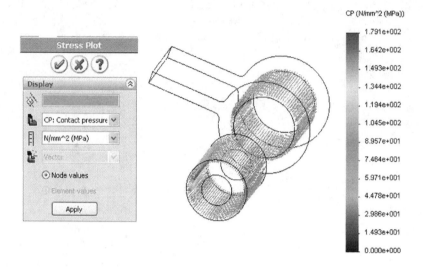

Figure 20-13: Contact stresses developed due to shrink fit

You may also wish to analyze radial stresses SX, which requires selecting the axis of the holes as reference geometry.

Connectors

Connectors are modeling tools used in defining connections between assembly components. COSMOSWorks offers four types of connectors: **Rigid, Spring, Pin, Elastic Support, Bolt, Spot weld** and **Link**. Their use is briefly explained in the following table. For more information refer to COSMOSWorks help which offers extensive explanations backed by examples.

TYPE OF CONNECTOR	FUNCTION
Rigid	Defines a rigid link between the selected faces. Faces connected by a rigid link do not translate or rotate in relation to each other.
Spring	Connects a face of a component to a face of another component by defining total stiffness or stiffness per area. Both normal and shear stiffness can be specified. The two faces must be planar and parallel to each other. The springs are introduced in the common area of the projection of one of the faces onto the other. You can specify a compressive or tensile preload for the spring connector.
Pin	A pin connects cylindrical faces of two components. Two options are available: No Translation: specifies a pin that prevents relative axial translation between the two cylindrical faces. No Rotation: specifies a pin that prevents relative rotation between the two cylindrical faces. Additionally, axial and/or rotational pin stiffness can be defined.
Elastic support	Defines an elastic foundation between the selected faces of a part or assembly and the ground. The faces do not have to be planar. A total or distributed spring stiffness may be defined in a normal and tangential direction to the affected face. Elastic support is used to simulate elastic foundations and shock absorbers. Elastic support is the only type of connector that can be defined both for part and assembly

TYPE OF CONNECTOR	FUNCTION
Bolt	Defines a bolt connector between two components. The bolt connector accounts for bolt pre-load. Configurations with and without a nut are available.
Spot weld	You can define spot welds to weld two solid faces or two shell faces. You should also define a No penetration contact condition between the two faces for proper modeling.
Link	The Link connector ties any two locations in the model by a rigid bar that is hinged at both ends. The distance between the two locations remains unchanged during deformation. The link connector is available for static, buckling, and frequency studies.

Connector definition is called by right-clicking the *Load/Restraints* folder and selecting **Connectors** (Figure 20-14).

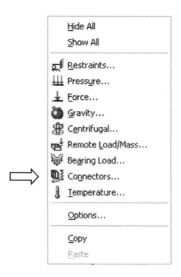

Figure 20-14: Connectors are called from the same pop up menu as Restraints and Loads

We will review the use of **Rigid** and **Pin** connectors using a model in the assembly CRANE. This model comes with defined **Rigid** connector and three **Pin** connectors.

The **Rigid** connector is shown in Figure 20-15.

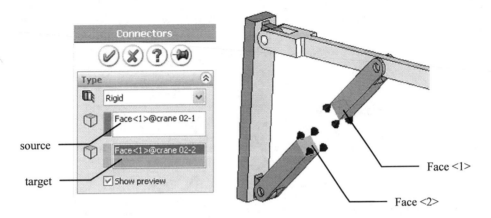

Figure 20-15: Rigid connector rigidly connects two faces

The selection of face as source and as target is arbitrary.

One of the **Pin** connectors is shown in Figure 20-16.

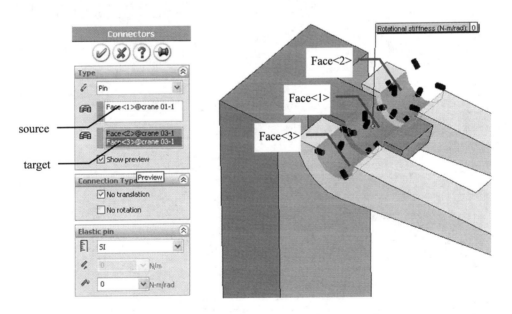

Figure 20-16: The Pin connector connects the middle face on one component to two faces of the other component

Note that torsional stiffness of the **Pin** connector shown in Figure 20-16 is specified as 0 (this is the default value). This means that **Pin** connectors allow for rotation between the two components. All degrees of freedom on the selected faces are coupled (must be the same) except for circumferential translations, which are disjoined.

To review the **Bolt** connector, open the assembly file FLANGE, which comes with six bolt connectors already defined. Right-click one of the *Bolt Connector* icons and select **Edit Definition** to open the **Connectors** windows (Figure 20-17).

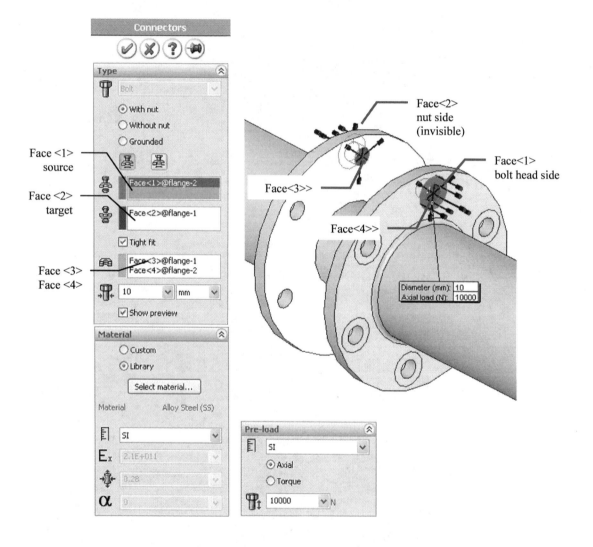

Figure 20-17 One of six bolt connectors in the FLANGE assembly model. Contact condition (no penetration) must be defined between touching faces of the two flanges. Tight fit must be defined to transmit the shear force in the absence of friction between flanges.

This illustration has been modified in a graphic program to show all entries in Connectors windows. Normally this would require scrolling.

Definition of **Bolt Connector** offers several options. In this example, we model the bolt with a nut. The bolt is made out of Alloy Steel, has tight fit, and a diameter of 10mm. The bolt is preloaded with an axial force 10,000N. All entries are shown in Figure 20-17.

Upon solution, review bolt forces available by right-clicking on the Results folder and selecting **List Pin/Bolt Force** (Figure 20-18).

Figure 20-18 Pop up menu activated by right-clicking the Results folder. Review bolt forces using the indicated selection.

Also investigate other results options in this menu.

Notes:

21: Implementation of FEA into the design process

Topics covered

- FEA driven design process
- FEA project management
- FEA project checkpoints
- FEA report

We have already stated that FEA should be implemented early in the design process and be executed concurrently with design activities in order to help make more efficient design decisions. This concurrent CAD-FEA process is illustrated in Figure 21-1.

Notice that design begins in CAD geometry and FEA begins in FEA-specific geometry. Every time FEA is used, the interface line is crossed twice: the first time when modifying CAD geometry to make it suitable for analysis with FEA, and the second time when implementing results.

This significant interfacing effort can be avoided if the new design is started and iterated in FEA-specific geometry. Only after performing a sufficient number of iterations do we switch to CAD geometry by adding all manufacturing specific features. This way, the interfacing effort is reduced to just one switch from FEA to CAD geometry as illustrated in Figure 21-1.

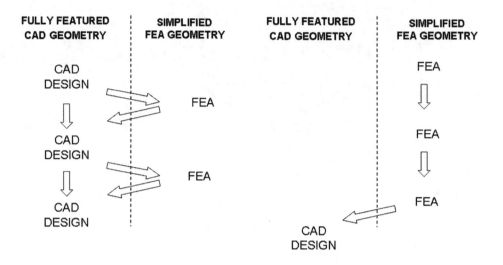

Figure 21-1: Concurrent CAD-FEA product development processes (left) and FEA driven product development process (right)

CAD-FEA design process is developed in CAD-specific geometry while FEA analysis is conducted in FEA-specific geometry. Interfacing between the two geometries requires substantial effort and is prone to error.

CAD-FEA interfacing efforts can be significantly reduced if the differences between CAD geometry and FEA geometry are recognized and the design process starts with FEA-specific geometry.

Let's discuss the steps in an FEA project from a managerial point of view. The steps in an FEA project that require the involvement of management are marked with an asterisk (*).

Do I really need FEA? *

This is the most fundamental question to address before any analysis starts. FEA is expensive to conduct and consumes significant company resources to produce results. Therefore, each use of FEA should be well justified.

Providing answers to the following questions may help to decide if FEA is worthwhile:

- Can I use previous test results or previous FEA results?
- Is this a standard design, in which case so no analysis is necessary?
- Are loads, supports, and material properties known well enough to make FEA worthwhile?
- Would a simplified analytical model do?
- Does my customer demand FEA?
- Do I have enough time to implement the results of the FEA?

Should the analysis be done in house or should it be contracted out? *

Conducting analysis in-house versus using an outside consultant has advantages and disadvantages. Consultants usually produce results faster while analysis performed in house is conducive to establishing company expertise leading to long-term savings.

The following list of questions may help in answering this question:

- How fast do I need to produce results?
- Do I have enough time and resources in-house to complete a FEA before design decisions must be made?
- Is in-house expertise available?
- Do I have software that my customer wants me to use?

Establish the scope of the analysis*

Having decided on the need to conduct FEA, we need to decide what type of analysis is required. The following is a list of questions that may help in defining the scope of analysis.

Is this project:

- A standard analysis of a new product from an established product line?
- The last check of a production-ready new design before final testing?
- A quick check of design in-progress to assist the designer?
- An aid to an R&D project (particular detail of a design, gauge, fixture etc.)?
- A conceptual analysis to support a design at an early stage of development (e.g., R&D project)?
- A simplified analysis (e.g., only a part of the structure) to help making a design decision?

Other questions to consider are:

- Is it possible to perform comparative analyses?
- What is the estimated number of model iterations, load cases, etc.?
- What are applicable criteria to evaluate results?
- How will I know whether the results can be trusted?

Establish a cost-effective modeling approach and define the mathematical model accordingly

Having established the scope of analysis, the FEA model must now be prepared. The best model is of course the simplest one that provides the required results with acceptable accuracy. Therefore, the modeling approach should minimize project cost and duration, but should account for the essential characteristics of the analyzed object.

We need to decide on acceptable idealization to geometry. This decision may involve simplification of CAD geometry by defeaturing, or idealization by using surface or wire frame representations. The goal is to produce a meshable geometry properly representing the analyzed problem.

Create a Finite Element model and solve it

The Finite Element model is created by discretization, or meshing, of a mathematical model. Although meshing implies that only geometry is discretized, discretization also affects loads and supports. Meshing and solving are both a largely automated step, but still require input, which depending on the software used, may include:

- Element type(s) to be used
- Default element size and size tolerance
- Definition of mesh controls (if any)
- Automesher type to be used
- Solver type and options to be used

Review results

FEA results must be critically reviewed prior to using them for making design decisions. This critical review includes:

- Verification of assumptions and assessment of results (an iterative step that may require several analysis loops to debug the model and to establish confidence in the results)
- Study of the overall mode of deformations and animation of displacements that loads and restraints have been defined properly
- Check for Rigid Body Motions

- Check for overall stress levels (at least the order of magnitude) using analytical methods in order to verify the applied loads
- Check for reaction forces and compare them with free body diagrams
- Review of discretization errors (e.g., by comparing nodal and element stresses)
- Analysis of stress concentrations and the ability of the mesh to model them properly
- Review of results in difficult-to-model locations, such as thin walls, high stress gradients, etc.
- Investigation of the impact of element distortions on the data of interest

Analyze results*

The exact execution of this step depends, of course, on the objective of the analysis.

- Present displacement results
- Present modal frequencies and associated modes of vibration (if applicable)
- Present stress results and corresponding factors of safety
- Consider modifications to the analyzed structure to eliminate excessive stresses and to improve material utilization and manufacturability
- Discuss results, and repeat iterations until an acceptable solution is found

Produce and accept report*

- Produce report summarizing the activities performed, including assumptions and conclusions
- Append the completed report with a backup of relevant electronic data

FEA project management requires the involvement of the manager during project execution. The correctness of FEA results cannot be established by only reviewing the analysis of the results. A list of progress checkpoints may help a manager stay in the loop and improve communication with the person performing the analysis. Several checkpoints are suggested in Figure 21-2. Following check points is especially recommended when junior engineers perform FEA.

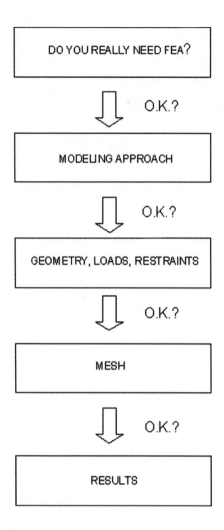

Figure 21-2: Checkpoints in an FEA project

Using the proposed checkpoints, the project is allowed to proceed only after the manager/supervisor has approved each step.

Even though each FEA project is unique, the structure of an FEA report follows similar patterns. The following are the major sections of a typical FEA report and their contents.

Section	Content
Executive Summary	Objective of the project, part/assembly number, project number, essential assumptions, results and conclusions, software used, information on where project backup is stored, etc.
Introduction	Description of the problem: Why did the project require FEA? What kind of FEA? (static, contact stress, frequency, etc.) What were the data of interest?
Geometry **Material** **Loads** **Restraints**	Description and justification of any simplification and/or idealization of geometry
	Justification of the modeling approach (e.g. solids, shells)
	Description of material properties and applicable failure criteria
	Description of loads and restraints, including load diagrams
Mesh	Description of the type of elements, global element size, any mesh control applied, number of elements, number of DOF, type of automesher used
	Justification of why this particular mesh is adequate to model the data of interest
Analysis of results	Presentation of displacement and stress results, including plots and animations
	Justification of the type of stress used to present results and failure criteria
	Discussion of errors in the results

Conclusions	Recommendations regarding structural integrity, necessary modifications, further studies needed
	Recommendations for follow-up testing procedure (e.g., strain-gauge test, fatigue life test)
	Recommendations on future similar analyses
Project documentation	Full documentation of design, design drawings, FEA model explanations, and computer back-ups
	Note that building in-house expertise requires very good documentation of the project besides the project report itself. Significant time should be allowed to prepare project documentation.
Follow-up	After completion of tests, append report with test results
	Discussion of correlation between analysis results and test results
	Discussion of corrective action taken in case correlation is unsatisfactory (may involve revised model and/or tests)

22: Glossary of terms

The following glossary provides definitions of terms used in this book.

Term	Definition
Adiabatic wall	An adiabatic wall is a wall where there is no heat going in or out; it is perfectly insulated. An adiabatic wall is one with no convection or radiation conditions defined.
Beam element	Beam element is intended for meshing wire frame geometry. Node of beam element has 6 degrees of freedom
Boundary Element Method	An alternative to the FEA method of solving field problems, where only the boundary of the solution domain needs to be discretized. Very efficient for analyzing compact 3D shapes, but difficult to use on more "spread out" shapes.
CAD	Computer Aided Design
Clean-up	Removing and/or repairing geometric features that would prevent the mesher from creating the mesh or would result in an incorrect mesh.
Constraints	Used in an optimization study, these are measures (e.g., stresses or displacements) that cannot be exceeded during the process of optimization. A typical constraint would be the maximum allowed stress magnitude.
Convergence criterion	Convergence criterion is a condition that must be satisfied in order for the convergence process to stop. In COSMOSWorks this applies to studies where h-Adaptive or the p-Adaptive solution has been selected.

Term	Definition
Convergence process	This is a process of systematic changes in the mesh in order to see how the data of interest change with the choice of the mesh and (hopefully) prove that the data of interest are not significantly dependent on the choice of discretization. A convergence process can be preformed as h-convergence or p-convergence. An h-convergence process is done by refining the mesh, i.e., by reducing the element size in the mesh and comparing the results before and after mesh refinement. Reduction of element size can be done globally, by refining mesh everywhere in the model, or locally, by using mesh controls to refine the mesh locally. An h-convergence analysis takes its name from the element characteristic dimension h, which changes from one iteration to the next. A p-convergence analysis does not affect element size, which stays the same throughout the entire convergence analysis process. Instead, element order is upgraded from one iteration to the next. A p-convergence analysis is done automatically in an iterative solution until the user-specified convergence criterion is satisfied. Sometimes, the desired accuracy cannot be achieved even with the highest available p-element order. In this case, the user has to refine the p-element mesh manually in a fashion similar to traditional h-convergence, and then re-run the iterative p-convergence solution. This is called a p-h convergence analysis.
Defeaturing	Defeaturing is the process of removing (or suppressing) geometric features in CAD geometry in order to simplify the finite element mesh or make meshing possible.
Design scenario	In COSMOSWorks, this is an automated analysis of sensitivity of selected results to changes of selected parameters defining the model.

Design variable	Used in an optimization study, this is a parameter (e.g. dimension) that we wish to change within a defined range in order to achieve the specified optimization goal.
Discretization	This defines the process of splitting up a continuous mathematical model into discrete "pieces" called elements. A visible effect of discretization is the finite element mesh. However, model mass, loads and restraints are also discretized.
Discretization error	This type of error affects FEA results because FEA works on an assembly of discrete elements (mesh) rather than on a continuous structure. The finer the finite element mesh, the lower the discretization error, but the solution takes more time.
Element stress	This refers to stresses at Gauss points of a given element. Stresses at different Gauss points are averaged amongst themselves (but not with stresses reported by other elements) and one value is assigned to the entire element. Element stresses produce a discontinuous stress distribution in the model.
Finite Difference Method	This is an alternative to the FEA method of solving a field problem, where the solution domain is discretized into a grid. The Finite Difference Method is generally less efficient for solving structural and thermal problems, but is often used in fluid dynamics problems.
Finite Element	Finite elements are the building blocks of a mesh, defined by position of their nodes and by functions approximating distribution of sought after quantities, such as displacements or temperatures.
Finite Volumes Method	This is an alternative to the FEA method of solving field problem, similar to the Finite Difference Method.
Frequency analysis	Also called modal analysis, a frequency analysis calculates the natural frequencies of a structure as the associated modes (shapes) of vibration. Modal analysis does not calculate displacements or stresses.

Gaussian points	These points are locations in the element where stresses are first calculated. Later, these stress results can be extrapolated to nodes.
h-adaptive solution	Iterative solution which involves mesh refinement. Iterations continue until convergence requirements are satisfied or the maximum number of iterations is reached.
h-element	An h-element is a fine element for which the order does not change during solution. Convergence analysis of the model using h-elements is done by refining the mesh and comparing results (like deflection, stress, etc.) before and after refinement. The name, *h-element*, comes from the element characteristic dimension *h*, which is reduced in consecutive mesh refinements.
Idealization	This refers to making simplifying assumptions in the process of creating a mathematical model of an analyzed structure. Idealization may involve geometry, material properties, loads and restraints.
Idealization error	This type of error results from the fact that analysis is conducted on an idealized model and not on a real-life object. Geometry, material properties, loads, and restraints all are idealized in models submitted to FEA.
Linear material	This is a type of material where stress is a linear function of strain.
Mesh diagnostic	This is a feature of COSMOSWorks that determines which geometric entities prevented meshing when meshing fails.
Meshing	This refers to the process of discretizing the model geometry. As a result of meshing, the originally continuous geometry is represented by an assembly of finite elements.
Modal analysis	See Frequency analysis.
Modeling error	See Idealization error.
Nodal stresses	These stresses are calculated at nodes by extrapolating stress from Gauss points and then averaging stresses (coming from different elements) at nodes. Nodal stresses are "smoothed out" and, by virtue of averaging, produce continuous stress distributions in the model.

Numerical error	This is round-of error accumulated by solver
Optimization goal	Also called an optimization objective or an optimization criterion, the optimization goal specifies the objective of an optimization analysis. For example in an optimization study, you could choose to minimize mass or maximize frequency.
p-element	P-elements are elements that do not have pre-defined order. Solution of a p-element model requires several iterations while element order is upgraded until the difference in user-specified measures (e.g., total strain energy, RMS stress) becomes less than the requested accuracy. The name *p-element* comes from the *p-order* of polynomial functions, which defines the displacement field in the element. This order is upgraded during the iterative solution.
p-adaptive solution	This refers to an option available for static analysis with solid elements only. If the p-Adaptive solution is selected (in the properties window of a static study), COSMOSWorks uses p-elements for an iterative solution. A p-adaptive solution provides results with narrowly specified accuracy
Pre-load	Pre-load is a load that modifies the stiffness of a structure. Pre-load may be important in a static or frequency analysis if it significantly changes structure stiffness.

Principal stress	Principal stress is the stress component that acts on the side of an imaginary stress cube in the absence of shear stresses. General 3D state of stress can be presented either by six stress components (normal stresses and shear stresses) expressed in an arbitrary coordinate system or by three principal stresses and three angles defining the cube orientation in relation to that coordinate system.
Rigid body mode or **Rigid body motion**	This refers to a mode of vibration with zero frequency found in structures that are not fully restrained or not restrained at all. A structure with no supports has six rigid body modes. Rigid body mode is the ability to move without elastic deformation. In the case of a fully supported structure, the only way the structure can move under load is by deforming its shape. If a structure is not fully supported, it can move as a rigid body without any deformation. Rigid body motions are only allowed in Frequency analysis.
RMS stress	Root Mean Square stress. RMS stress may be used as a convergence criterion if the p-adaptive solution method is used.
Sensitivity study	See Design scenario.
Shell element	Shell elements are intended for meshing surfaces. The shell element that is used in COSMOSWorks is a triangular shell element. Triangular shell elements have three corner nodes. If this is a second order triangular element, it also has mid-side nodes, making the total number of nodes equal to six. Each node of a shell element has 6 degrees of freedom.

Small Displacement assumption	Analysis based on small displacements assumes that displacements caused by loads are small enough to not significantly change structure stiffness. Analysis based on this assumption of small deformations is also called linear geometry analysis or small displacement analysis.
	However, the magnitude of displacements itself is not the deciding factor in determining whether or not those deformations are indeed small. What matters is whether or not those displacements significantly change the stiffness of the analyzed structure.
Steady state thermal analysis	Steady state thermal analysis assumes that heat flow has stabilized and no longer changes with time.
Structural stiffness	Structural stiffness is a function of shape, material properties, and restraints. Stiffness characterizes structural response to an applied load.
Symmetry boundary conditions	These refer to displacement conditions defined on a flat model boundary allowing only for in-plane displacement and restricting any out-of-plane displacement components. Symmetry boundary conditions are very useful for reducing model size if model geometry, load, and supports are all symmetric.
Tetrahedral solid element	This is a type of element used for meshing solid models. A tetrahedral element has four triangular faces and four corner nodes. If used as a second order element (high quality in COSMOSWorks terminology) it also has mid-side nodes, making then the total number of nodes equal to 10. Each node of a tetrahedral element has 3 degrees of freedom.
Thermal analysis	Thermal analysis finds temperature distribution, temperature gradient and heat flux in a structure.

Transient thermal analysis	Transient thermal analysis is an option in a thermal analysis. It calculates temperature, temperature gradient, and heat flow changes over time as a result of time-dependent thermal loads and thermal boundary conditions.
Ultimate strength	The maximum stress that may occur in a structure. If the ultimate strength is exceeded, failure will take place (the part will break). Ultimate strength is usually much higher than yield strength.
Von Mises stress	This is a stress measure that takes into consideration all six stress-components of a 3D state of stress. Von Mises stress, also called Huber stress, is a very convenient and popular way of presenting FEA results because it is a scalar, non-negative value and because the magnitude of von Mises stress can be used to determine safety factors for materials exhibiting elasto-plastic properties, such as most types of steel.
Yield strength	The maximum stress that can be allowed in a model before plastic deformation takes place.

23: Resources available to FEA users

Many sources of FEA expertise are available to users. Sources include, but are not limited to:

- Engineering textbooks
- Software manuals
- Engineering journals
- Professional development courses
- FEA users' groups and e-mail exploders
- Government organizations

Readers of this book may wish to review the book "Finite Element Analysis for Design Engineers" which expands on many topics we discussed in this book (Figure 23-1).

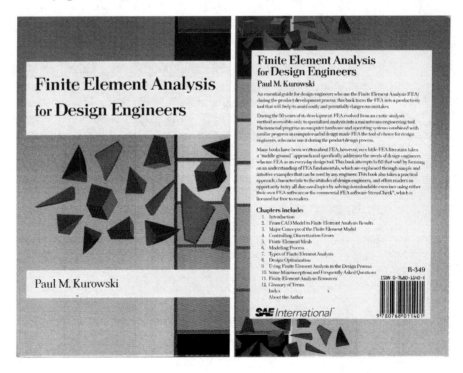

Figure 23-1 "Finite Element Analysis for Design Engineers" book

"Finite Element Analysis for Design Engineers" is available through the Society of Automotive Engineers (www.sae.org).

Engineering literature offers a large selection of FEA-related books, a few of which are listed here.

- Adams V., Askenazi A. "Building Better Products with Finite Element Analysis", Onword Press, 1998.

- Macneal R. "Finite Elements: Their Design and Performance", Marcel Dekker, Inc., 1994.

- Spyrakos C. "Finite Element Modeling in Engineering Practice", West Virginia University Printing Services, 1994.

- Szabo B., Babuska I. "Finite Element Analysis", John Wiley & Sons, Inc., 1991.

- Zienkiewicz O., Taylor R. "The Finite Element Method", McGraw-Hill Book Company, 1989.

Several professional organizations like the Society of Automotive Engineers (SAE) and the American Society of Mechanical Engineers (ASME) offer professional development courses in the field of the Finite Element Analysis. More information on FEA related courses offered by the SAE and ASME can be found on www.sae.org and www.asme.org.

With so many applications for FEA, various levels of importance of analysis, and various FEA software, attempts have been made to standardize FEA practices and create a governing body overlooking FEA standards and practices. One of leading organizations in this field is the National Agency for Finite Element Methods and Standards, better know by its acronym NAFEMS. It was founded in the United Kingdom in 1983 with a specific objective "To promote the safe and reliable use of finite element and related technology." NAFEMS has published many FEA handbooks like:

- A Finite Element Primer

- A Finite Element Dynamics Primer

- Guidelines to Finite Element Practice

- Background to Benchmarks

The full list of these excellent publications can be found on www.nafems.org.

A valuable source of information on different types of analysis and analysis techniques can be found on the COSMOS Companion site. The COSMOS Companion is a series of short subject presentations provided by SolidWorks at the following address:

http://www.cosmosm.com/pages/news/COSMOS_Companion.html

Another internet site with a number of FEA related publications is presented by Design Generator, Inc. Publications related to FEA fundamentals, training and implementation can be found at:

http://www.designgenerator.com/publications.htm